Legal Information

© 2024
Author and Editor: M.Eng. Johannes Wild
Author Reference: A94689H39927F
Email: 3dtech@gmx.de

The complete imprint of the book can be found on the last pages!

Table of contents

Chapter 1 | Introduction

1.1 What to expect in this course and what you will learn

Welcome to the "FreeCAD" advanced Course — Part 2. Thank you for choosing this book! With this course, you can take your CAD skills to the next level and become a professional.

Please note: This book is the sequel to the book "FreeCAD | Design Projects" (ISBN: 9783987421020) and the basic book "FreeCAD | Step by Step" (ISBN: 9783987420924). If you are a beginner in "FreeCAD", you should start with these two books to learn all the basics first. Further details about these titles can be found at the end of this book.

In this course, we will dive into creating complex assemblies, working with surfaces, threads, lettering, and will get to know other advanced features. Four complex design projects will help you refine your CAD skills and become an expert in designing with "FreeCAD".

We will start with a brief review of essential settings before jumping straight into our first design project. This course is very hands-on and therefore does not have separate theory chapters — you will gain enough theoretical knowledge while working on the projects.

As an engineer with many years of experience in CAD, my goal is to make advanced "FreeCAD" design easy to understand. Throughout the book, I use several teaching methods to ensure you gain a thorough understanding of the software and design principles.

In summary, you can learn:

- To deepen the basics of "FreeCAD"
- Get to know new 2D and 3D features
- Handling surfaces and curves
- Creating lettering and threads
- New approaches and techniques in design
- How to create complex individual parts and assemblies
- Independent work following technical instructions / drawings

Design projects:

- Scissors with a handle,

- Hole punch with attachments,
- Computer mouse with a scroll wheel,
- Dumbbell bar with weight plates.

"FreeCAD" is open-source software and therefore available free of charge. Just go to the official website https://www.freecadweb.org and download the latest version.

Get your copy of this book now and start improving your skills in "FreeCAD" today!

1.2 "FreeCAD" Basic settings

Before we start with the design projects, we first compare the program settings. After starting the program, click on the "Edit" button ① and select the option "Preferences ..." ②. Note: The positions of the settings listed below may change between versions. If you cannot find the settings in your version immediately, it is easiest to browse through the menus a little and look for the terms.

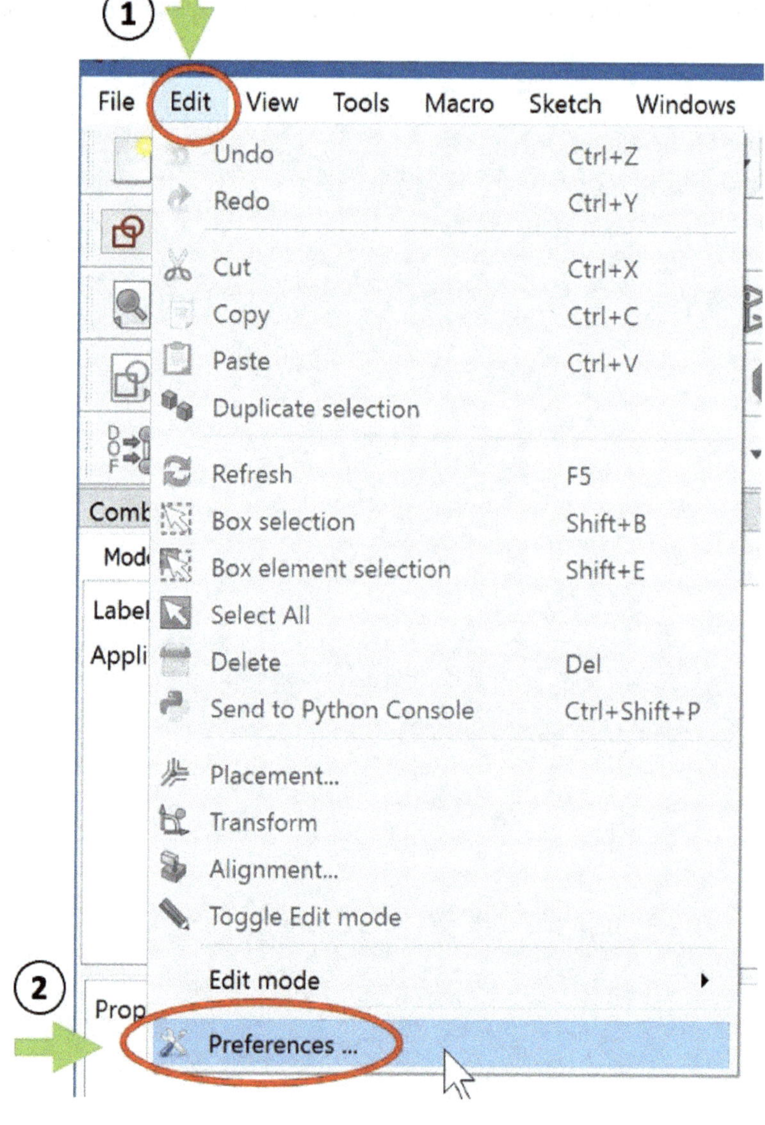

"FreeCAD" automatically selects the language of your operating system when it is started for the first time. You can change this setting in the "General" section (① and ②). We check if the language is set to English ③. Another important setting in the "General" section is the preferred unit system. We use the standard units "Standard (mm/kg/s/degree)" ④. We can also change the appearance of the display in the "General" tab. If this is not important to you, you can simply leave the default setting "No style sheet" ⑤ here. In this section, we can also set the size of the icons for the toolbar commands. It is best to use the setting "Medium (24px)" ⑥ if this has not yet been selected. In addition, please check whether the setting "Tree view mode" is set to "Combo View" ⑦.

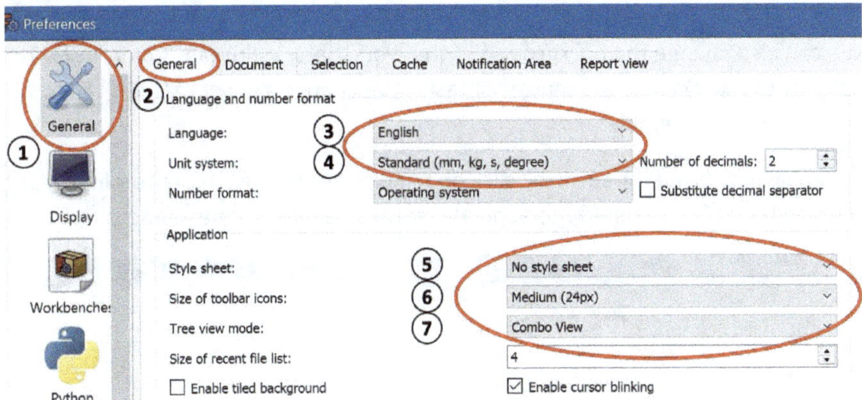

We then need to check whether the coordinate system is activated in the "Display" ① section and in the "3D View" ② tab. The option "Show coordinate system in the corner" ③ must be checked. We can also increase the display size of the orbit cube ("navigation cube") and the coordinate system here (④, ⑤ and ⑥).

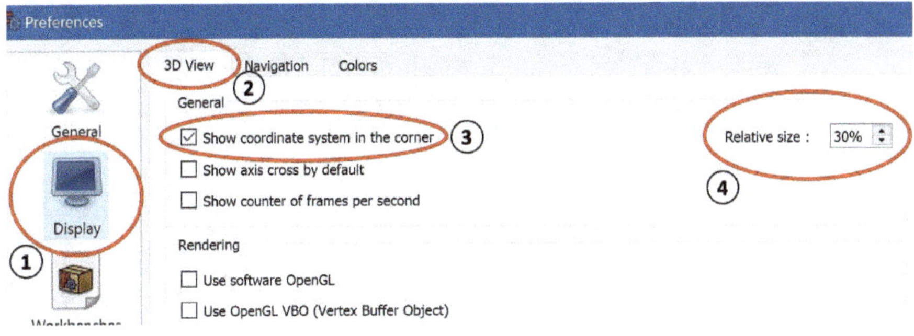

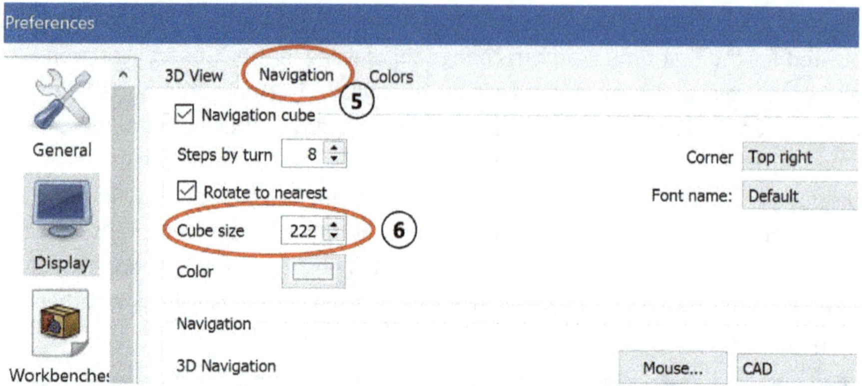

We can also change the background of the workspace in the tab "Colors" ⑦. The color of the workspace is a matter of taste. We change the background ⑧ to the color white, for example.

If you have changed the settings, click on "Apply" at the bottom of the window and then on the "OK" button ⑨ to apply the settings and close the window.

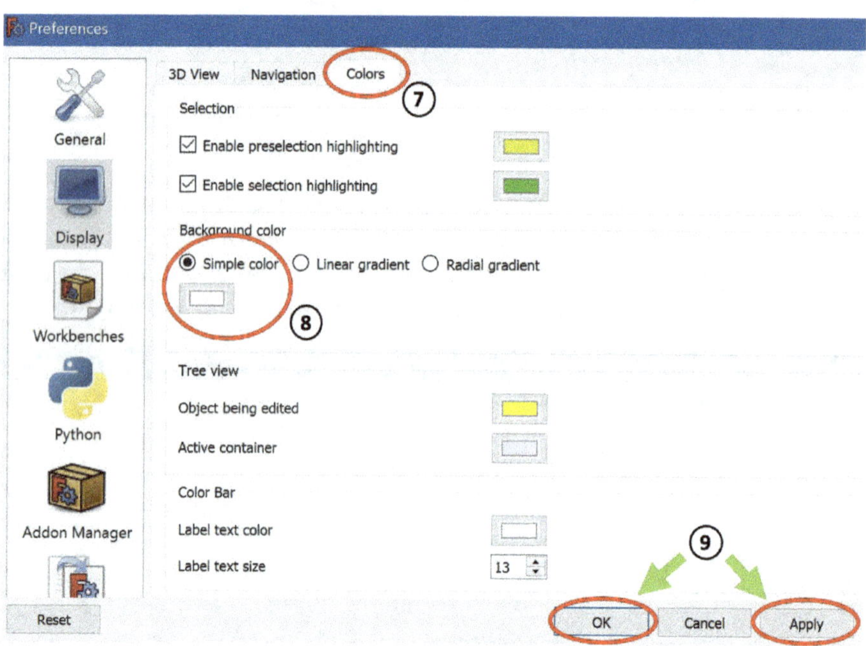

We also need to make a few important basic settings in the "Sketcher" workspace. As you probably remember, we usually create the 2D sketches of our 3D objects in the "Sketcher" workspace.

To make the settings, we must first switch to this workspace (① and ②) and then open the settings again ("Preferences ..." / ③ and ④).

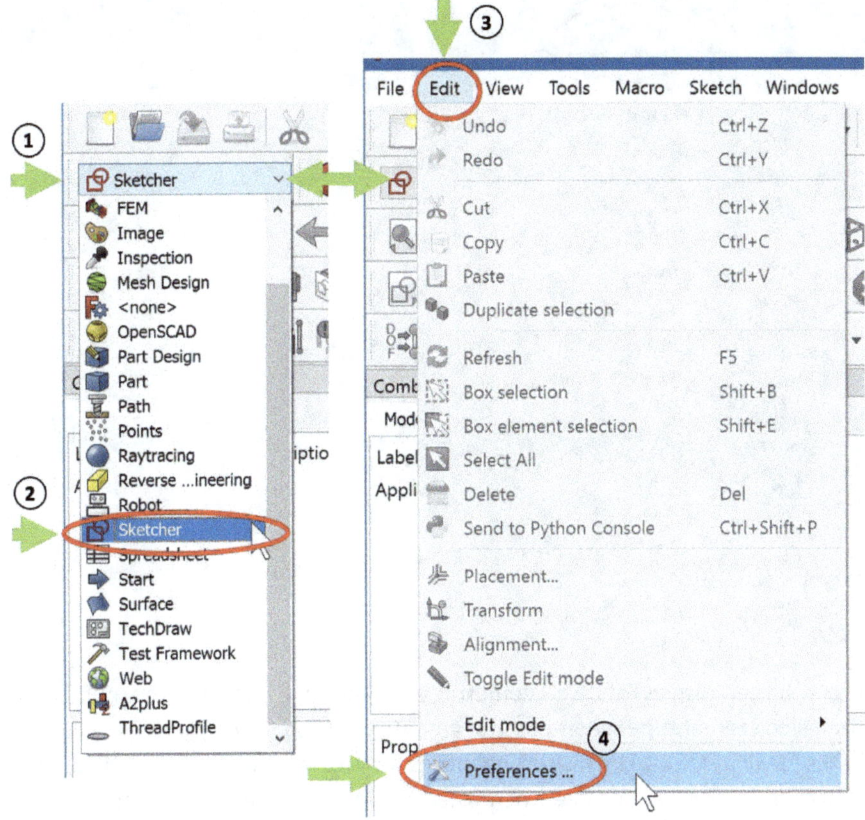

Here we make a few settings regarding the display. To do this, we navigate to the "Sketcher" section in the sidebar of the window ①.

You could — if desired — activate the drawing grid for 2D sketch creation in the tab "Grid" ② by ticking the option "Grid" ③. The setting "Grid Auto Spacing" would cause the grid size to change automatically based on the size of the view. This option could also be activated here if required.

As the grid would tend to get in the way when sketching freely, we will deactivate it for now. This will make it easier for you to follow the drawn lines.

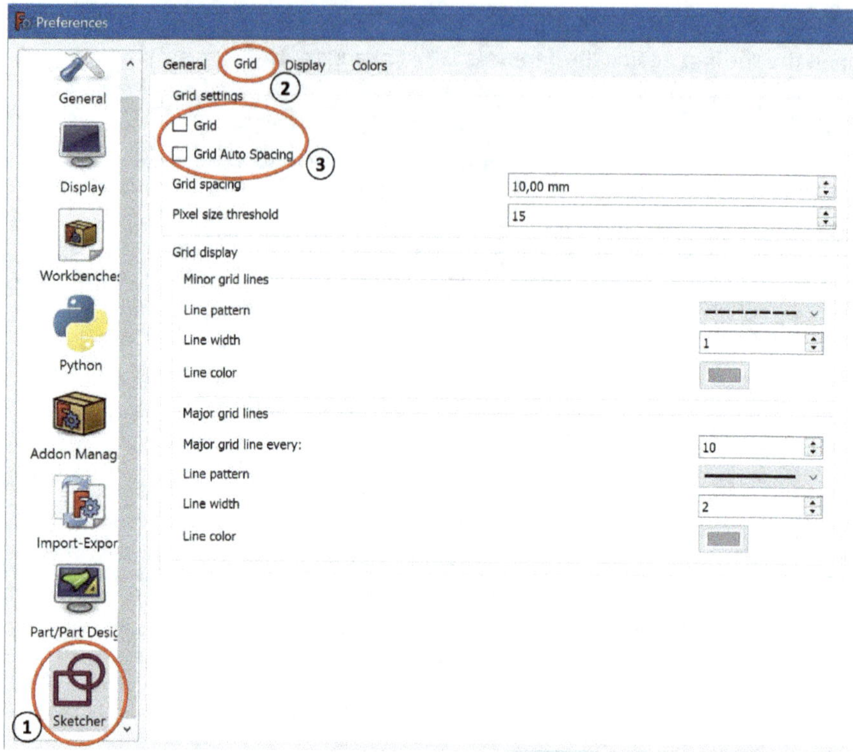

For a better display of the geometric elements, we increase the number of displayed segments per geometry ("Segments per geometry") ⑤ to the value 500 in the "Display" tab ④. This setting makes a circle, for example, appear rounder. However, it is only noticeable when you zoom in very close to the geometry.

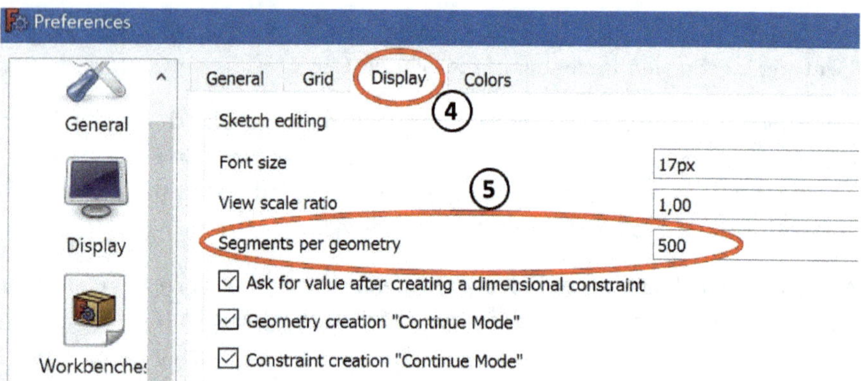

Furthermore, please set the selection fields ⑦ - ⑨ in the "Colors" tab ⑥ to the color black or a similarly dark color if you had selected a white or light background.

Otherwise, we would not be able to recognize the geometric elements later. Now we can close the preferences.

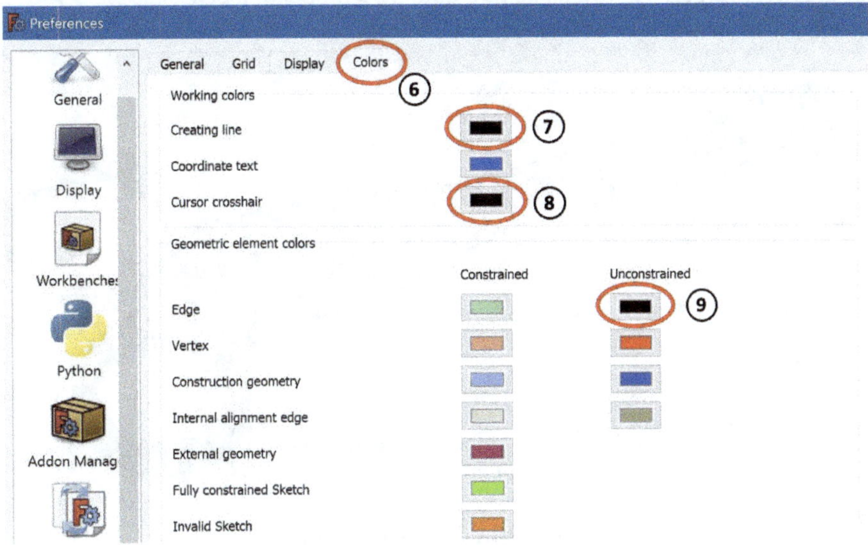

How you can move the drawing area, and the 3D object, depends on the preselected mode. You can select this at the bottom right of the drawing area ①. It is best to select the mode "CAD" ②. Navigation then takes place as shown ③. "Pan" means move. "Rotate", "Zoom" and "Select" should be clear.

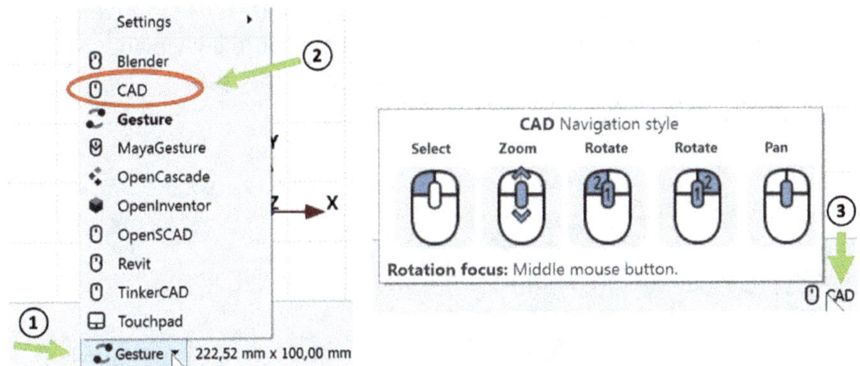

Now all our settings are identical, and therefore no problems should occur in the further course. We can now get started with the design projects!

Chapter 2 | Project 1: Scissors with handle

In the first project, we would like to create a 3D model of a pair of scissors with a handle. This should look as follows.

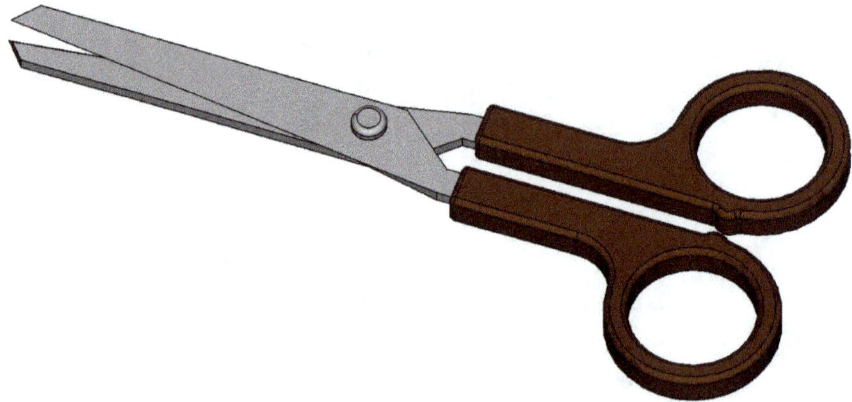

2.1 The scissor blade

We start with the design of the scissor blade, the geometry of which consists of a cutting area, a handle area and a hole for assembly. As an advanced user, you should already know that we switch to the "Part Design" ① workspace at the start of each project. There we create a new document using the command "New" ② and a body using the command "Create body" ③. We then need to create a 2D sketch. For the first scissor blade, we create a sketch on the x-y plane using the command "Create Sketch" ④ and by selecting the plane. The program automatically takes us to the "Sketcher" workspace.

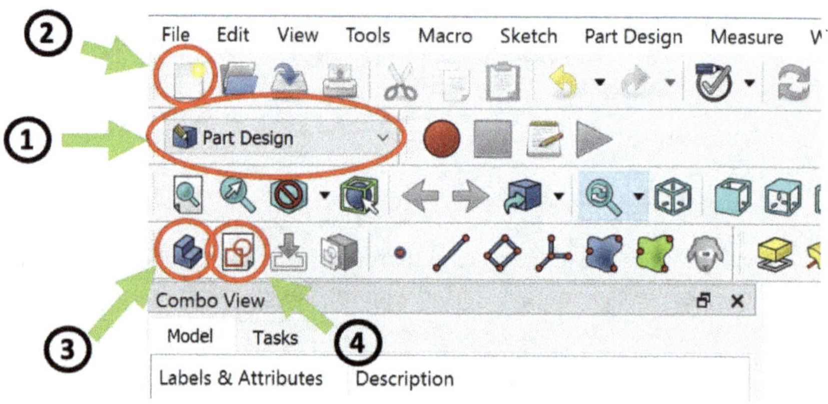

In this sketch, we draw the first two lines with the command "Create line" ①. For the first line, we start at the coordinate origin ② and draw horizontally with a length of 65 mm ③ to the left. We then create a diagonal line that should extend from the origin ② to the top right. Dimension this with 10 mm and 9 mm (④ and ⑤). The horizontal and vertical dimensions can be set using the commands "Constrain horizontal distance" and "Constrain vertical distance" ⑥. Note: The green vertical line and the red horizontal line represent the x and y axes.

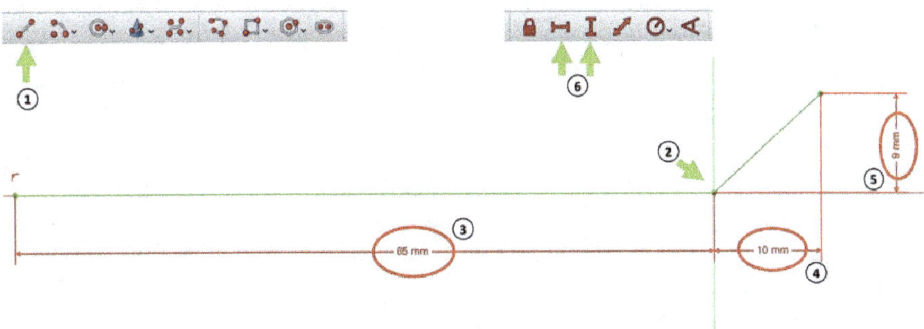

Next, we create the hole for the bolt that will later hold the scissors together. To do this, we use the command "Create Circle" ① and place the circle at the desired position ②. We can define a 4 mm diameter with the command "Constrain diameter" ③. We must also add a 5 mm distance between the center of the circle and the origin of the coordinates in both horizontal and vertical directions.

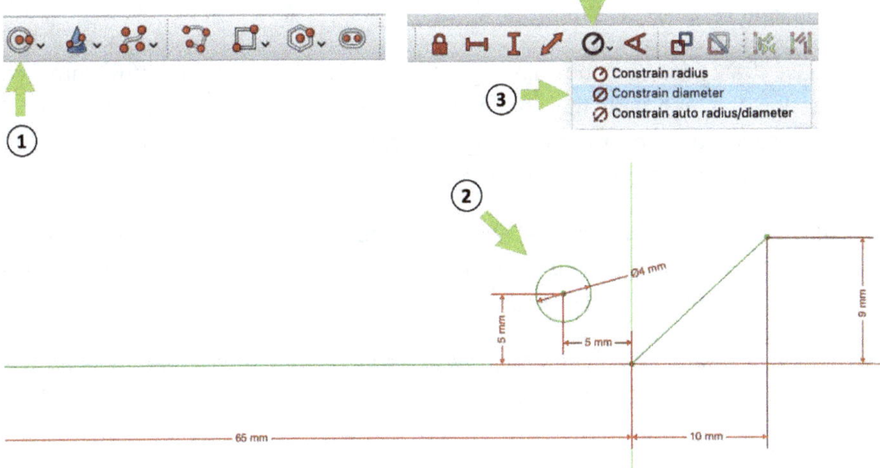

At the back, we add a 20 mm horizontal line (① and ④) to the sketch, followed by a short 5 mm vertical line ②. We then create another horizontal line ③ and set the length of this line to be identical to the first horizontal line drawn. We do this with the condition "Constrain equal" ⑤. First select the condition and then click on both lines (③ and ④) one after the other. We will attach the handle to this part of the scissor blade later.

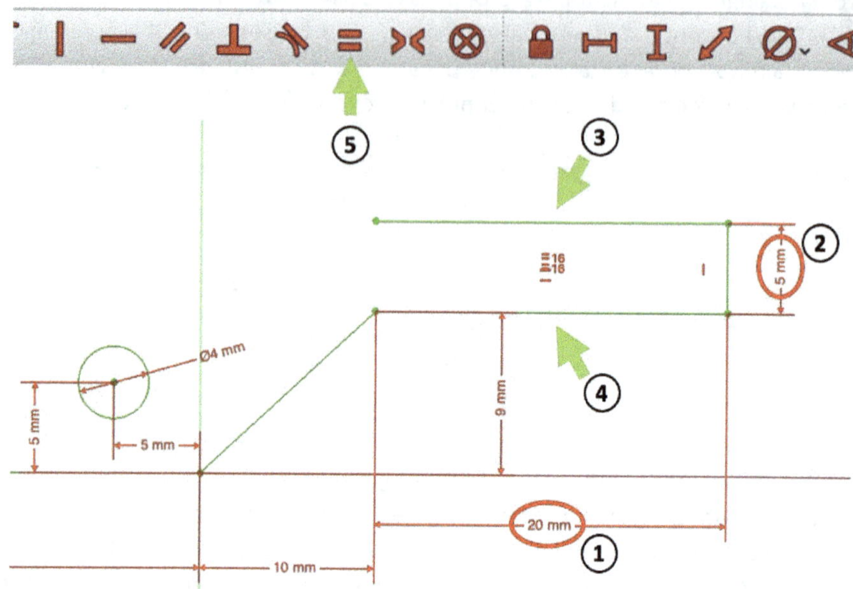

To complete the sketch of the scissor blade, let's add the tip. For this, we draw a short diagonal line ① in the left-hand area, which we dimension horizontally at 2.5 mm and vertically at 3 mm.

Next, we draw a long diagonal line ② with a horizontal dimension of 60 mm. We dimension this line with the command "Constrain angle" ③ at an angle of 7 degrees to the lower horizontal line ④. To achieve this, we click on both lines after activating the command and enter the value.

Finally, we complete the geometry with another diagonal line ⑤. This line does not need to be dimensioned as the length and angle of inclination are automatically determined by the rest of the geometry.

Once all elements have been correctly positioned and dimensioned, we end the sketch by clicking on "Close" in the "Tasks" tab in the "combo view" on the left-hand side of the window <u>or</u> alternatively by clicking on the ESC key.

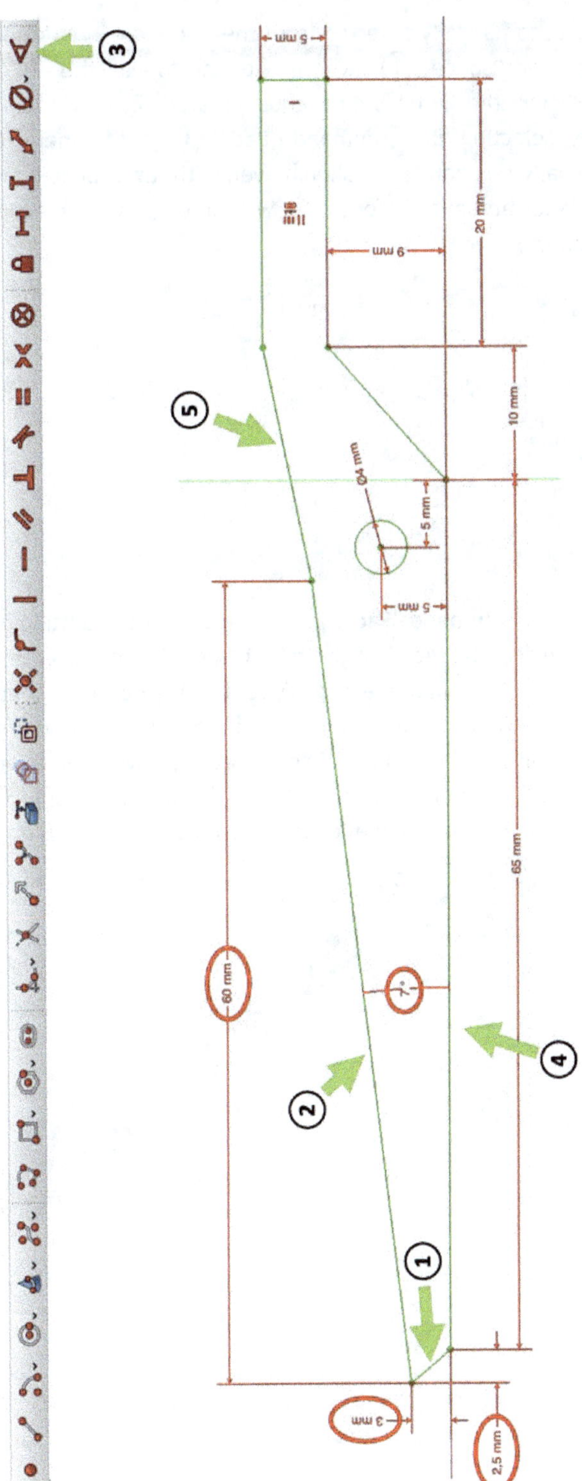

5 mm
9 mm
20 mm
10 mm
Ø4 mm
5 mm
5 mm
60 mm
65 mm
3 mm
2,5 mm

The software automatically takes us from the "Sketcher" workspace back to the "Part Design" workspace ①, where we can make the 2D sketch three-dimensional. Select the sketch we have just created ② in the "combo view" and extrude it by selecting the command "Pad" ③ in the menu bar. However, "FreeCAD" usually recognizes the sketch even without prior selection. If we want to show all planes and axes beforehand, we can do this by selecting "Origin" ④ and clicking on the space bar.

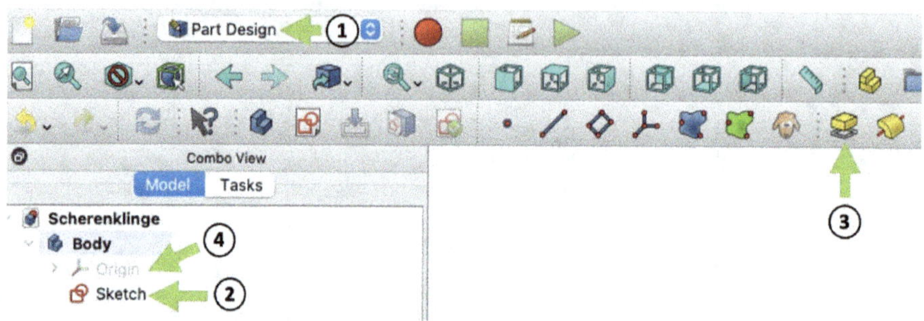

After selecting the command "Pad", we are shown various setting options for this command in the "combo view" in the tab "Tasks" ①. In this case, we select the option "Two dimensions" in the setting "Type" ②. We do this so that the 3D profile will be extruded symmetrically above and below the x-y plane (the x-y plane intersects the profile in the middle). To complete this, we need to enter a value of 1 mm for "Length" ③ and "2nd length" ④. We can then confirm with "OK" to execute the command. The total thickness of the scissor blade is now 2 mm.

In the last step, we will create the cutting edge of the scissor blade. For this, we create an angled surface with the command "Draft" ①. After selecting the command, click on the side surface ② of the scissors and then enter an angle of 36° in the "Draft angle" ③ field. We then have to click on the button "Neutral plane" ④ and select the upper surface of the scissors ⑤.

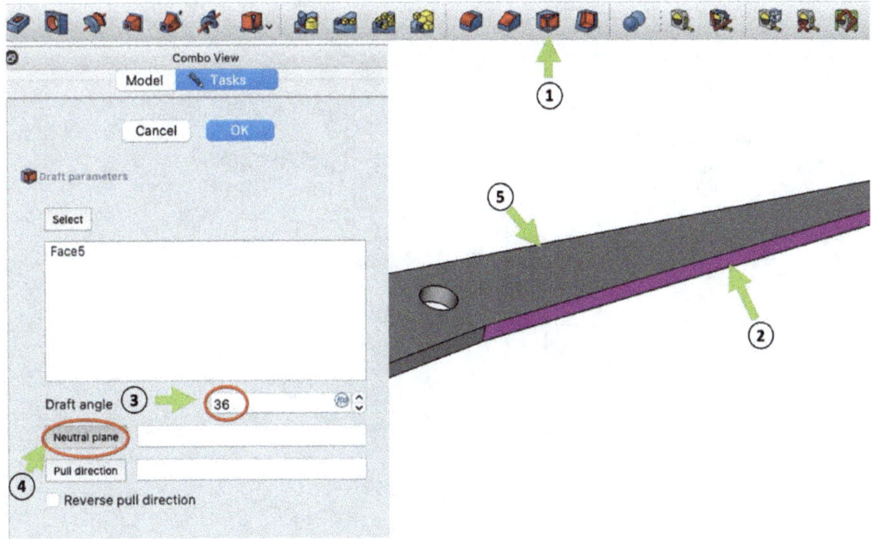

The cutting edge is now created as shown.

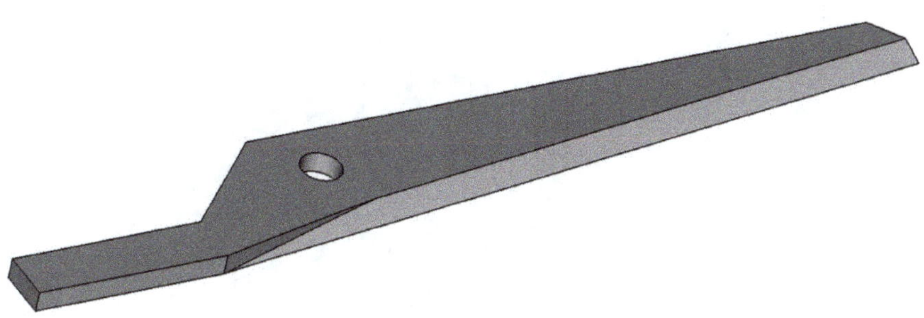

Once we have saved this first component of our later assembly, we can dedicate ourselves to creating the handle of the scissors in a new document. The model of the scissors is symmetrical, so we can simply use the scissor blade — and the handle — twice. For this reason, we only need to design one part of each.

2.2 The handle of the scissors

To create the handle part, we first create a body as usual in the "Part Design" workspace using the "Create body" command and start a sketch on the x-y plane (see section 2.1). Now we can sketch the following profile. You are welcome to try drawing the profile on your own first. This is a good exercise. Don't worry, the individual solution steps will follow later.

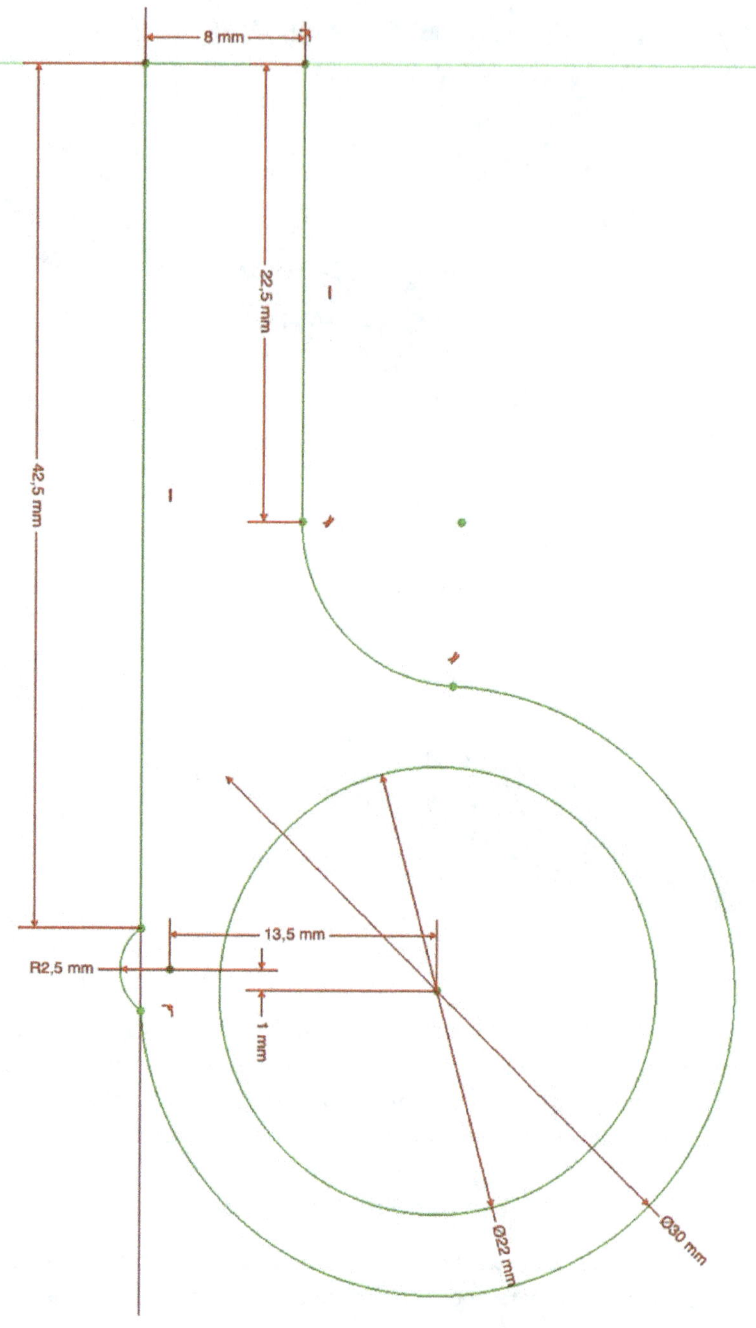

We start with an 8 mm vertical line ①, the starting point of which should be at the coordinate origin ②. In the upper area, we connect a 22.5 mm horizontal line ③. Finally, we draw a 42.5 mm line ④, which should be congruent with the x-axis.

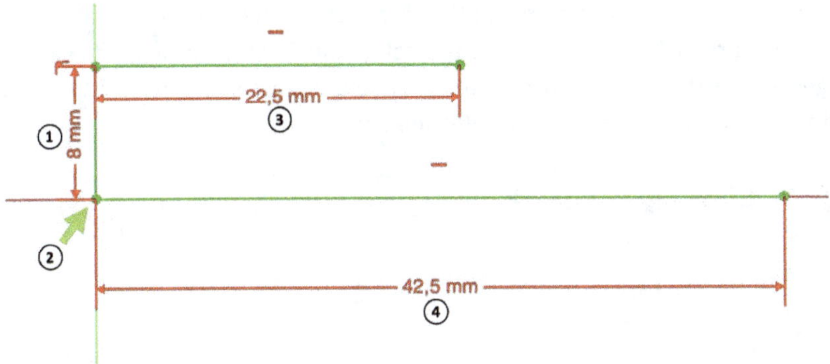

Next, we create two 3-point arcs. We do this with the command "End points and rim point" ②. The first arc should connect to the 22.5 mm line at point ③ and end at point ④. The second arc should connect to the 42.5 mm long line at point ⑤ and end at point ⑥ on the x-axis. It is sufficient to position the end points of the arcs approximately. We will do the final positioning in a later step.

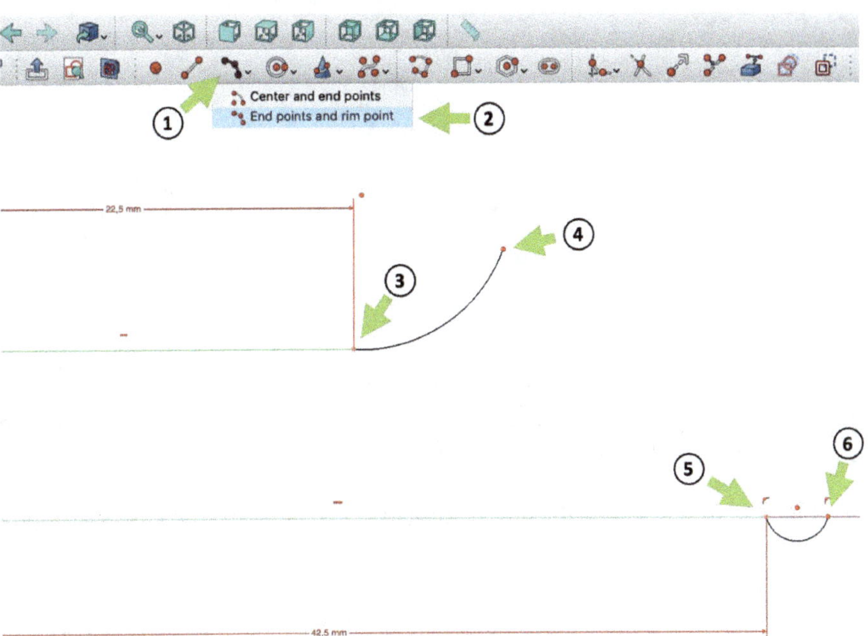

Note: If an error message appears, you can simply click it away. This generally applies to all steps. The steps shown here should still always work. In most cases, the error messages only appear because the correct settings have not yet been selected at the start of a command.

We then create another 3-point arc that connects to the previous arc at point ① and ends at point ②. We then use the command "Constrain tangent" ③ to set tangential constraints. Simply click on the two arc segments in point ①. We also do this in point ④ with the horizontal line and the arc.

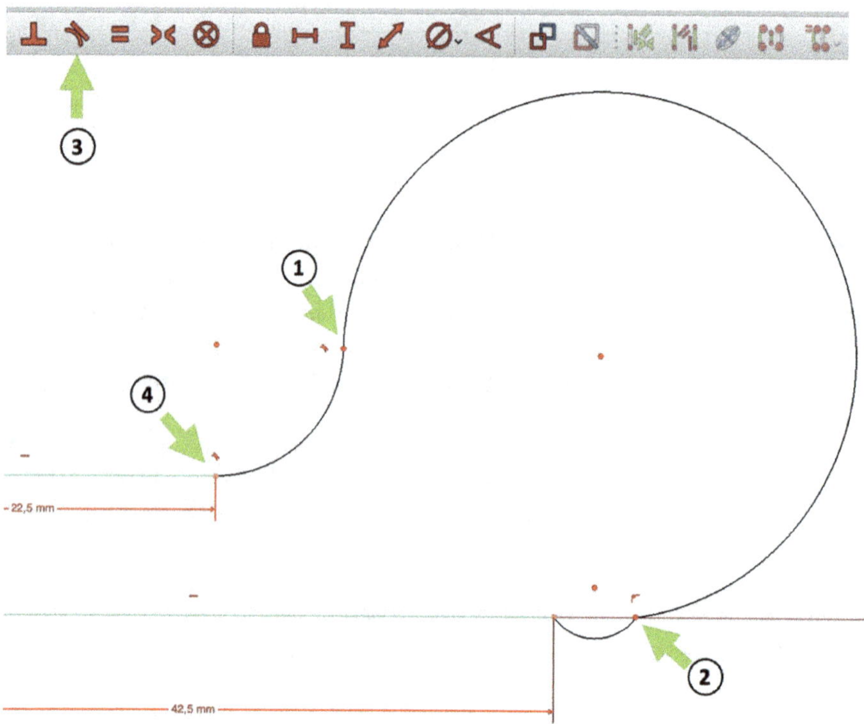

For a full definition of the sketch, we need to add a few dimensions. Firstly, we dimension the small 3-point arc with a radius of 2.5 mm ① and secondly the large 3-point arc with a diameter of 30 mm ②.

We then define the horizontal and vertical distance between the two arc centers (③ and ④) by assigning a horizontal dimension of 1 mm and a vertical dimension of 13.5 mm.

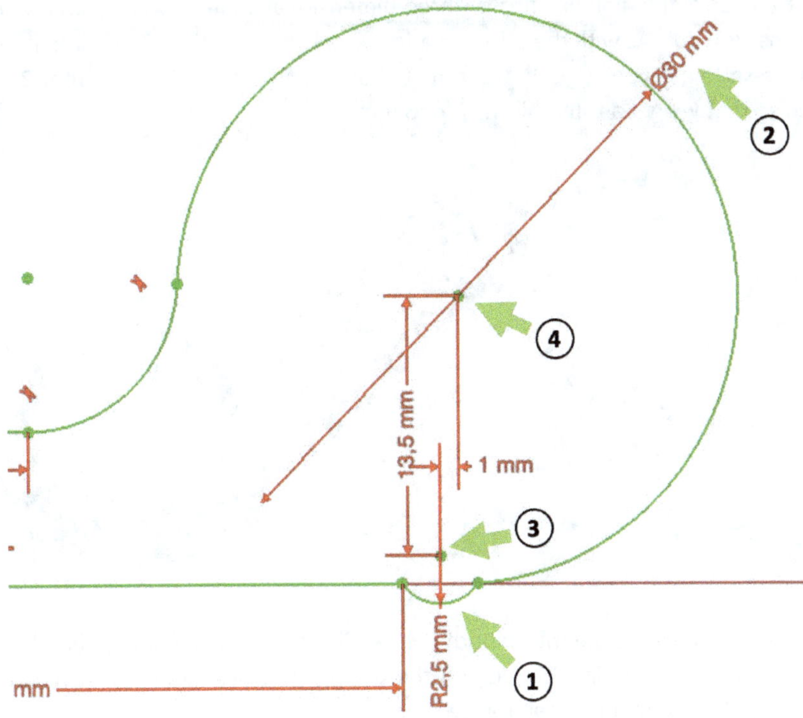

Finally, we add a 22 mm circle (concentric to the arc) to complete the sketch and then end the sketch with the ESC key.

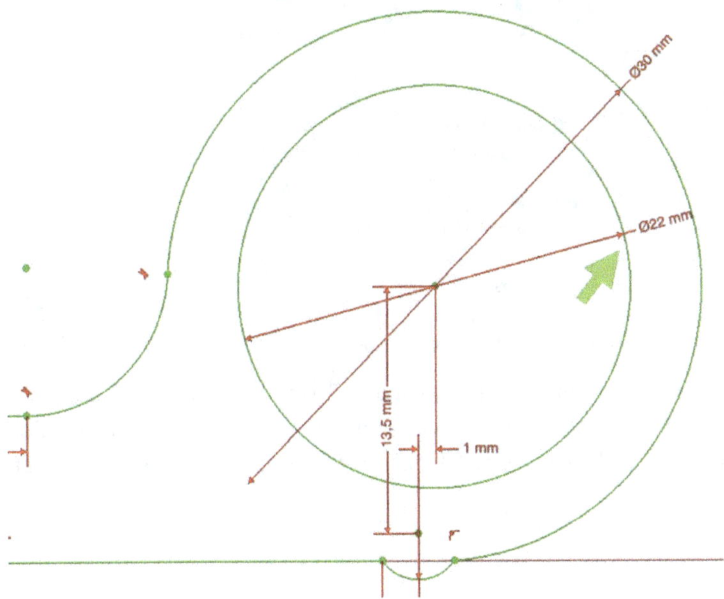

To transform the handle into a three-dimensional object, we again use the command "Pad". As with the scissor blade, we select the option "Two dimensions" for the setting "Type" ①. At "Length" ② and "2nd length" ③ we enter 3 mm each so that we get a total thickness of 6 mm.

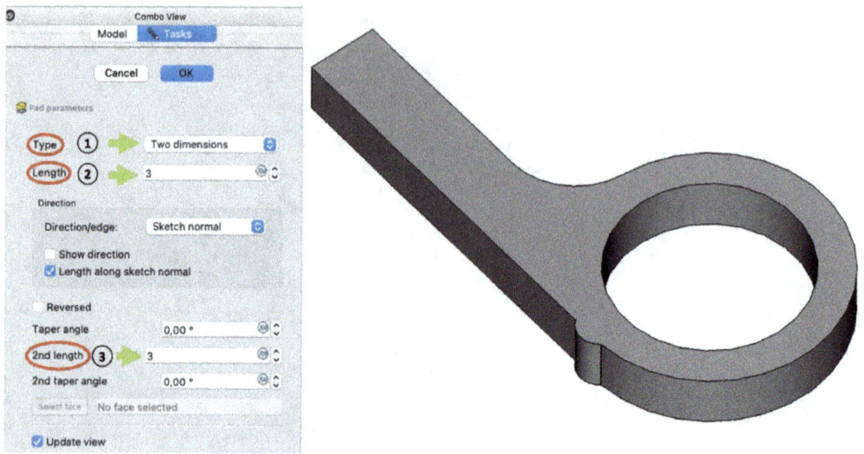

Next, we want to change the color of the handle to give the design a refined look. This can be done by right-clicking on the body ① in the combination view and selecting the command "Appearance" ②.

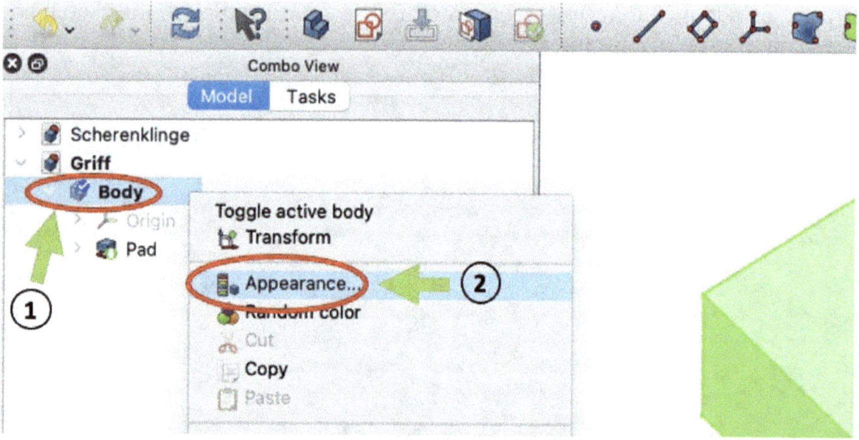

This opens the command's settings in the "Tasks" tab ① in the "combo view". We now need to click on the color selection button ② for the "Shape color" option and then select a brown shade ③, which we can change as desired using the slider ④.

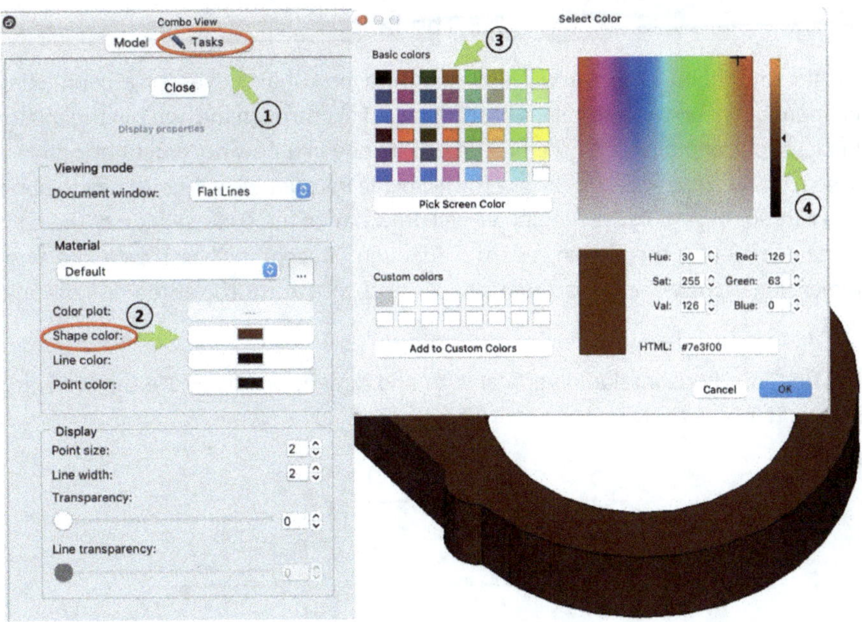

Finally, we make a few fillets so that the handle is more comfortable to hold. We create these with the command "Fillet" ① from the menu bar. In the "Tasks" ② tab, we can set a radius ③ of 1 mm and activate the "Use All Edges" ④ option, as we want to round all edges of the part. Then simply confirm with "OK" ⑤ and the handle is done.

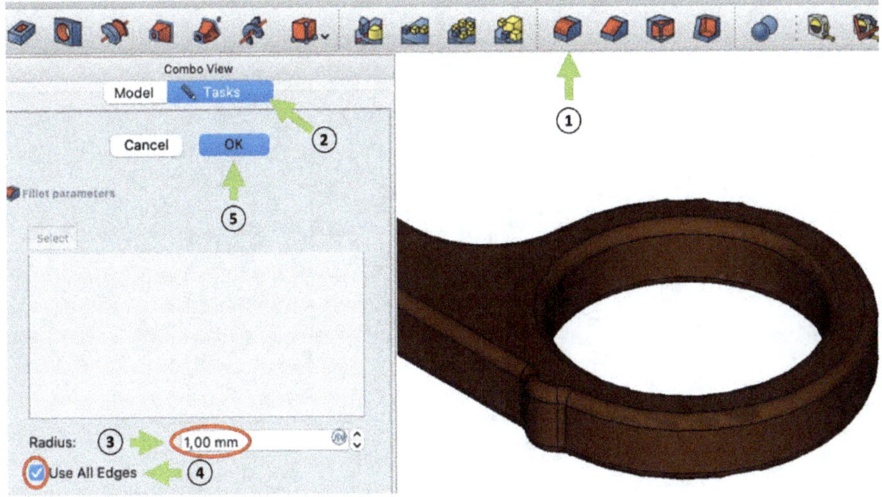

Excellent! Now, all we need is the scissor bolt and can then assemble the individual parts to form the scissors.

2.3 The bolt

As already announced, we will now create a bolt that serves as a connecting element between the two halves of scissors and fits through the holes in the scissor blades. To accomplish this, we first create a body in a new document and then a sketch on the x-z plane. We choose the x-z plane because we will construct the bolt as a rotational part. In the sketch, we will draw half of the cross-section of the bolt. After completing the sketch, use the function "Revolution" to rotate the two-dimensional cross-section around an axis and thus form the three-dimensional bolt.

But first, use horizontal and vertical lines and dimensions to create the following sketch. It is best to start at the coordinate origin ①.

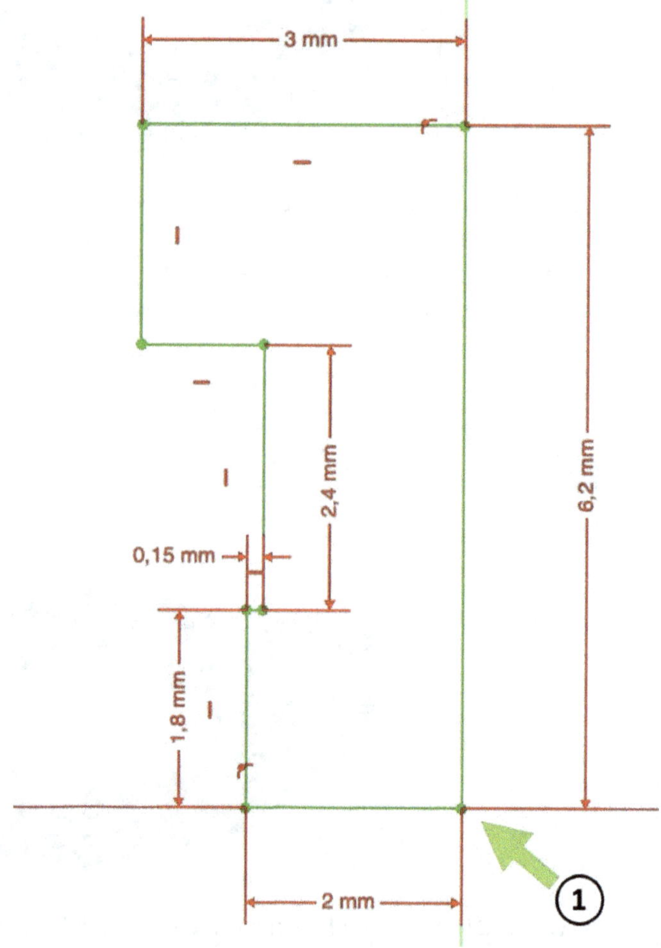

After we have closed the sketch with the ESC key, we activate the command "Revolution" ①. Normally, "FreeCAD" automatically recognizes the correct rotation axis and creates the following 3D model. If not, you must adjust the settings ②.

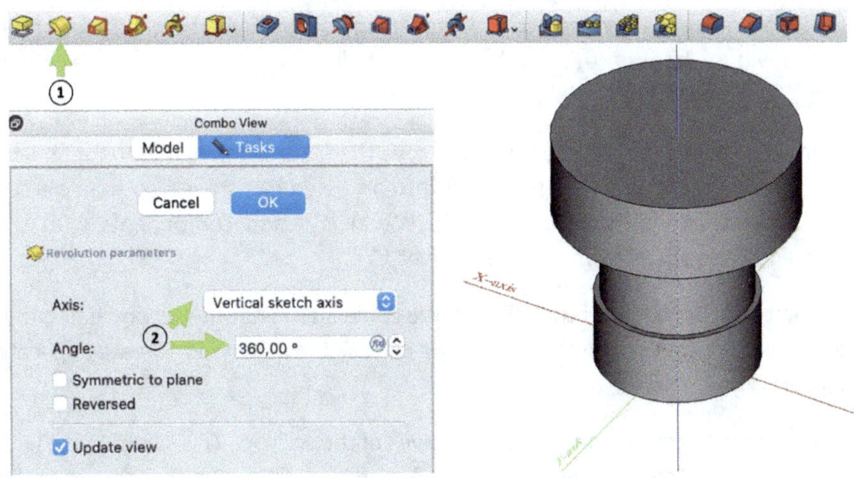

Now we need a rounding of the upper edge and a chamfer on the lower edge. The chamfer makes it easier to insert the bolt into the hole. For the rounding, we use the command "Fillet" ①, for the chamfer the command "Chamfer" ②. The fillet radius should be 1 mm, the chamfer should have a dimension of 0.3 mm. Simply select the command, click on the edge, and then enter the value.

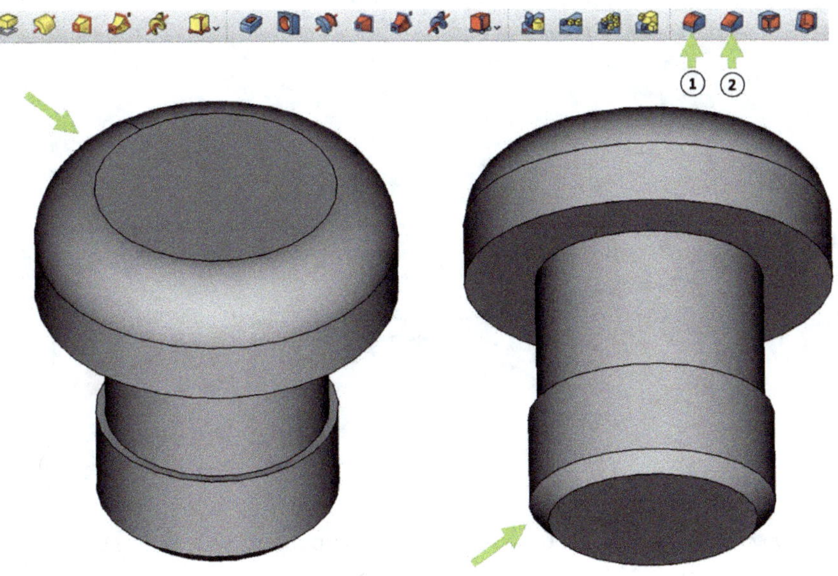

2.4 Assembling the scissors

Now that we have created all the individual parts, in this chapter we will take care of assembling the scissors. To do this, we will create a new document in the "A2plus" workspace. The "A2plus" workspace is intended for the assembly of individual parts and may need to be installed on your system. If you have not read the previous books, this website will help you with the installation: https://wiki.freecad.org/A2plus_Workbench

Before we can get started, we must first save the file under a name of our choice. The first individual part that we insert into an assembly is fixed so that we can mount all other individual parts on it. We always add components with the command "Add a part from an external file" ①.

Let's first insert one half of the scissors. We could also insert the second half of the scissors in the same way. Alternatively, we can use the command "Create duplicate of a part" ②.

Now we need to rotate one of the two halves of the scissors. This can be done with the command "Move the selected part" ③ and by rotating the component at the red point ④ (rotation around x-axis). We rotate by 180° so that the scissor halves are parallel to each other after the rotation.

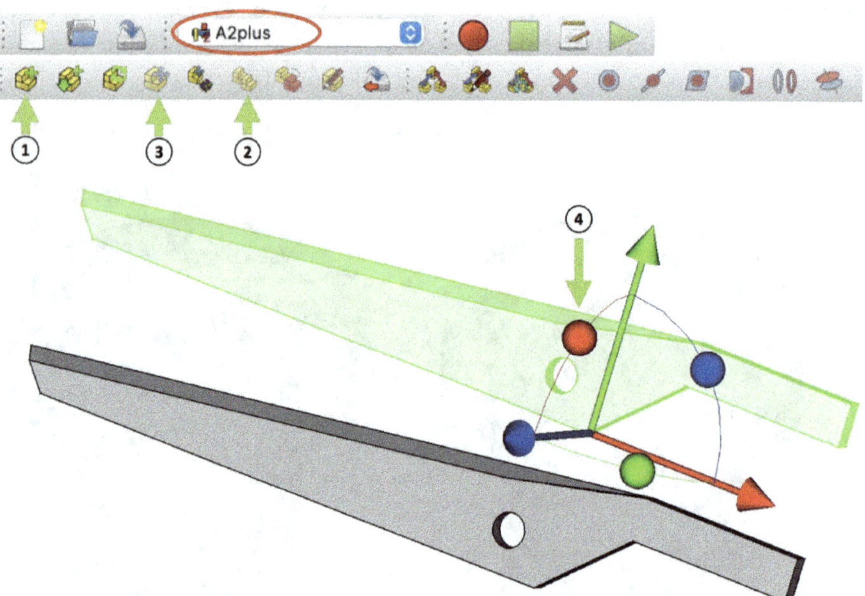

Then we create the first link between the two components. We do this by selecting the inner surfaces (① and ②) of the two holes by holding down the CTRL key and

then clicking on the command "Add AxisCoincident Constraint" ③. A window opens in which we can create the link by clicking on the button "Accept" ④.

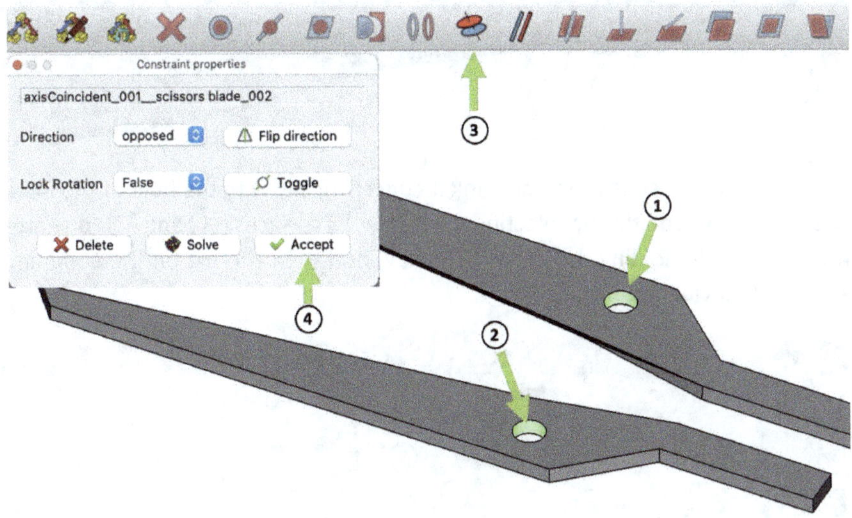

We need a second link for the two scissor halves. To do this, we select one upper and one lower side of the scissor halves by holding down the CTRL key (① and ②) and then click on the command "Add PlaneCoincident Constraint" ③.

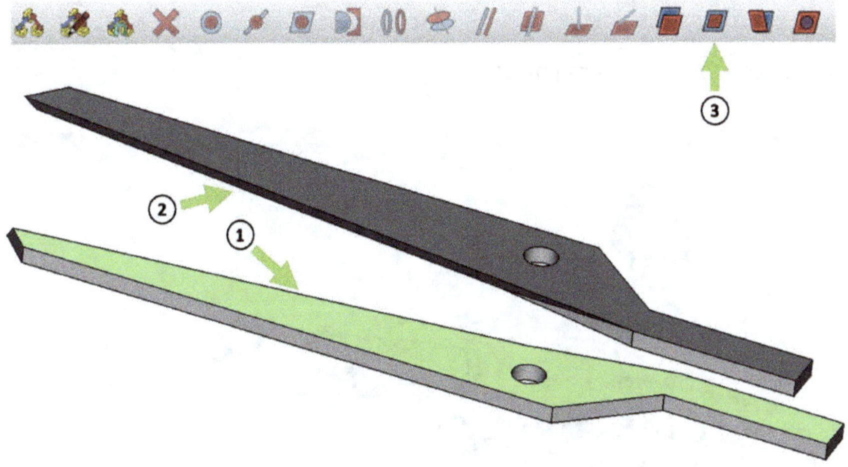

This should give us the following result. The scissor halves are opposite each other and the holes are congruent.

Now we insert the scissor bolt and link it concentrically with the hole of the lower scissor blade. We do this by selecting the displayed surfaces (①) and (②) one after the other (while holding down the CTRL key) and then clicking on the command "Add AxisCoincident Constraint" (③).

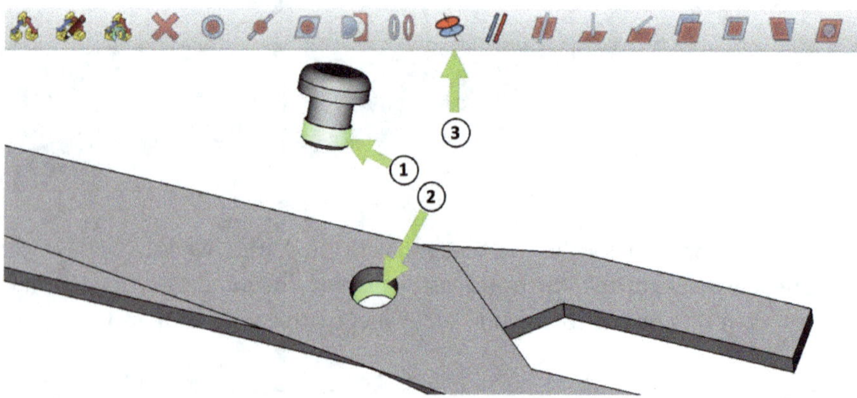

We also create a congruent link as shown.

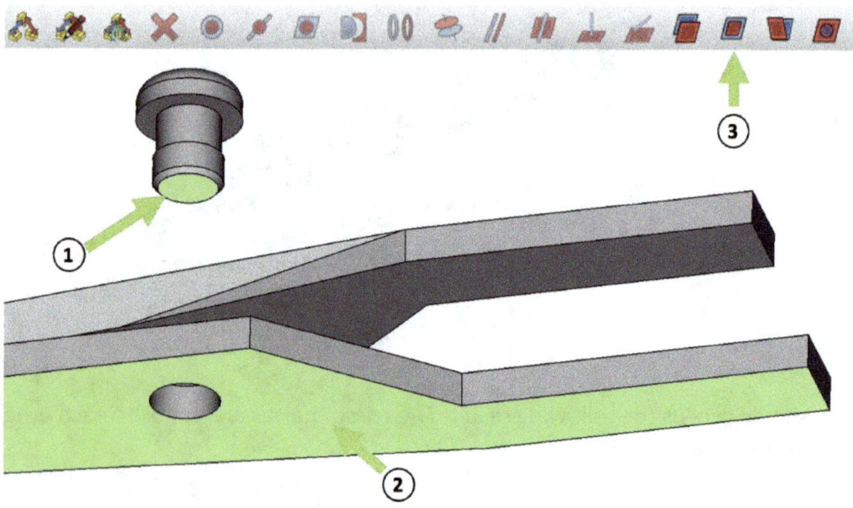

Now that the scissor bolt has been positioned correctly, we can mount the two handles. First, we insert one of the two handles and create a congruent link between the two end faces ① and ②. The selection is made — as usual — by holding down the CTRL key. Once we have clicked on the command "Add PlaneCoincident Constraint" ③, we must enter a distance of 17.5 mm in the settings window for the option "Offset" ④ so that the handle is positioned correctly.

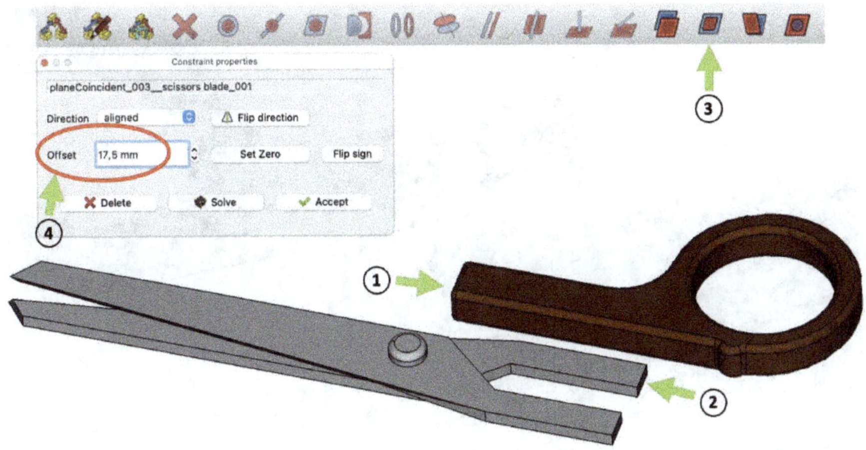

We create two further links in a similar way to fully define the position of the handle in 3D space. First, we link the two cover surfaces ① and ② and enter an offset of 3 mm ④.

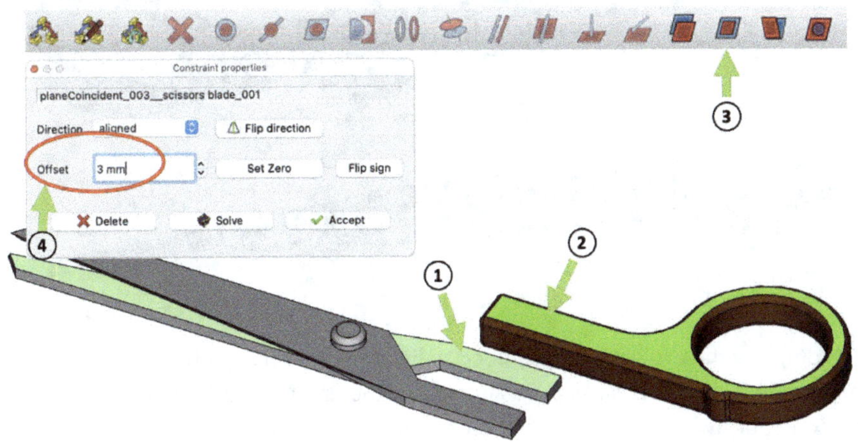

We then link the lateral surfaces ① and ② and enter an offset of -1.5 mm ④.

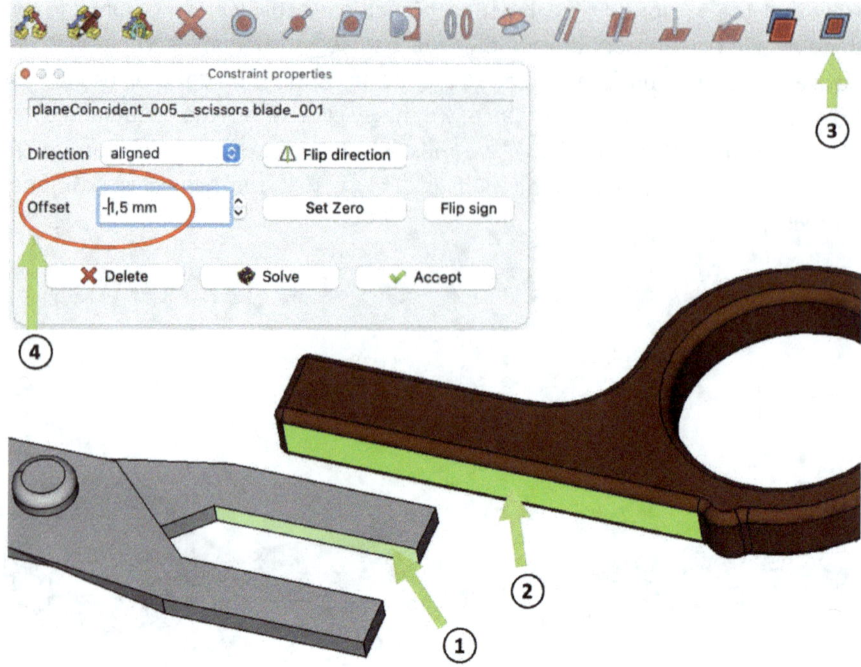

The handle should then be positioned as follows.

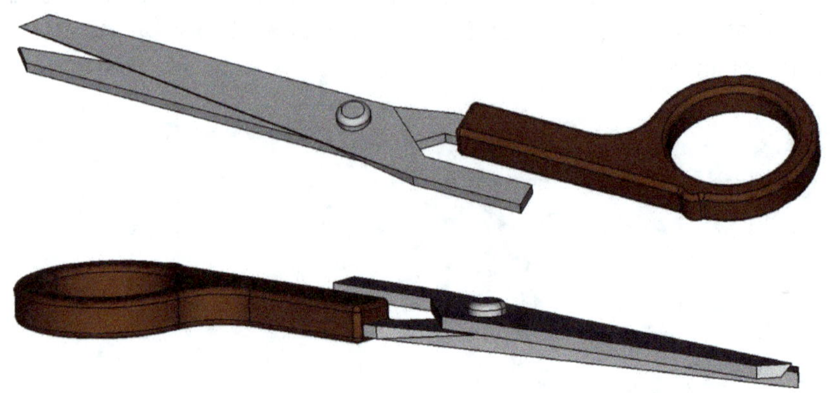

To complete the project, we only need to link the other handle to the scissors. The procedure for this is basically the same as for the first handle. You can certainly do this on your own. However, there is one difference, which I will explain below. If you link the two surfaces ① and ② together, you need an offset of - 1 mm (or +1 mm, depending on the position), as the handles must be positioned asymmetrically, as shown in detail ⑤. This is necessary so that the handles lie in one plane.

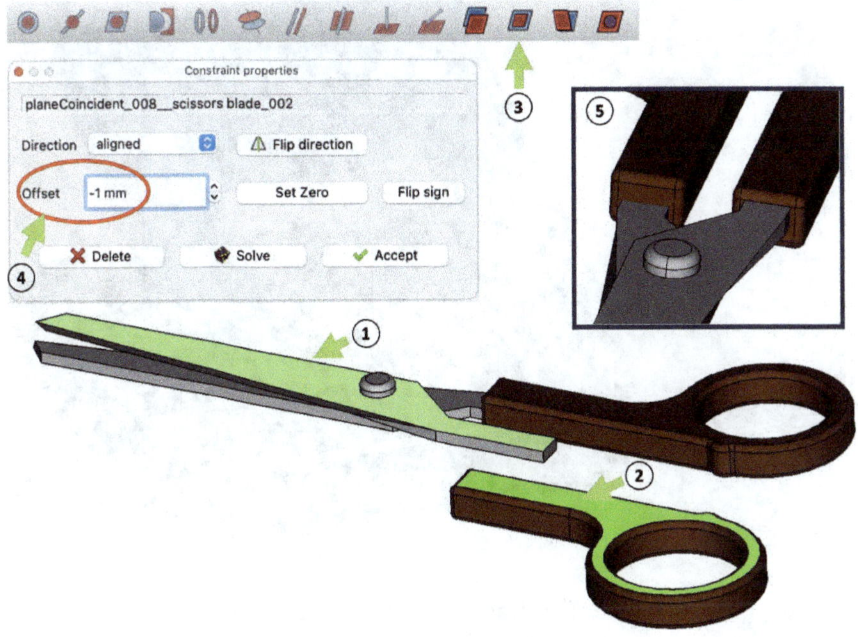

Perfect, now the 3D model of the scissors is done! Great achievement!

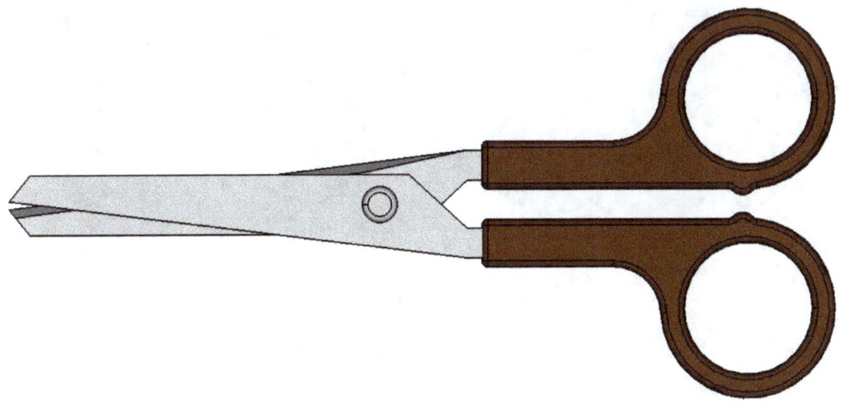

Once we have saved the project, we can move on to the next 3D model!

Chapter 3 | Project 2: Hole punch with attachments

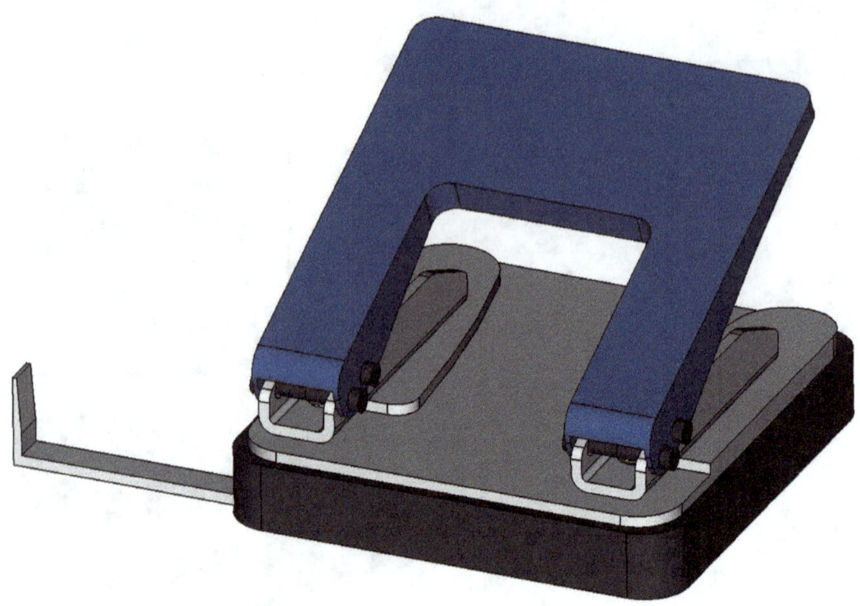

3.1 The base of the hole punch

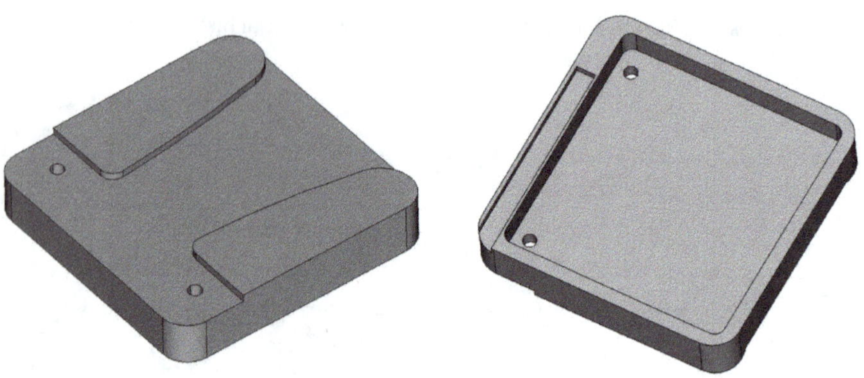

For the base of the hole punch, we create a new document and switch to the "Part Design" workspace. Here we can use the command "Create body" to create a body and the command "Create Sketch" to create a sketch on the x-y plane. The basic body consists of a square with 200 mm sides, the center of which lies at the coordinate origin ②. To achieve this as easily as possible, we use the command "Centered rectangle" ①.

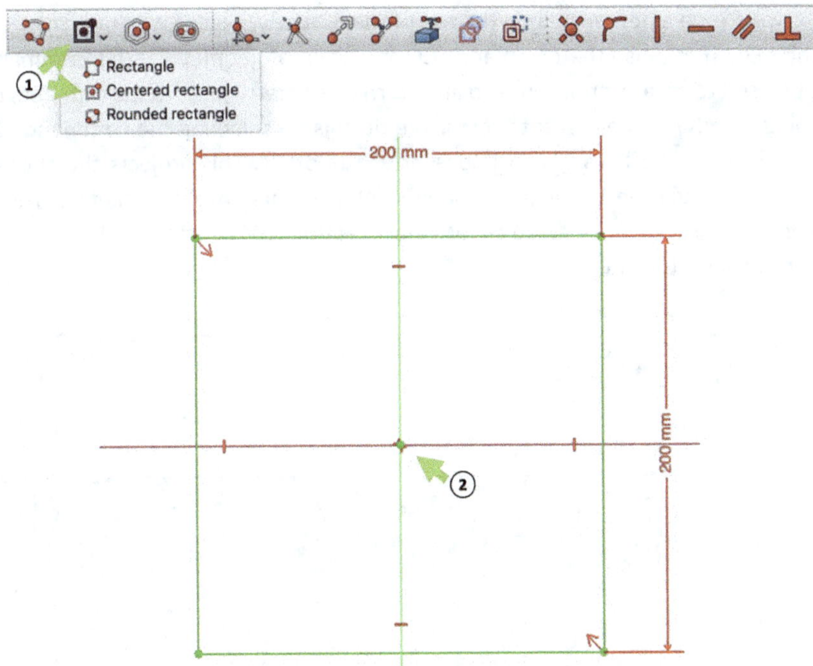

Then end the sketch with the ESC key and create a 30 mm high 3D body using the command "Pad" ①. Let's round the four vertical edges of the cuboid with the command "Fillet" ② and a radius of 20 mm.

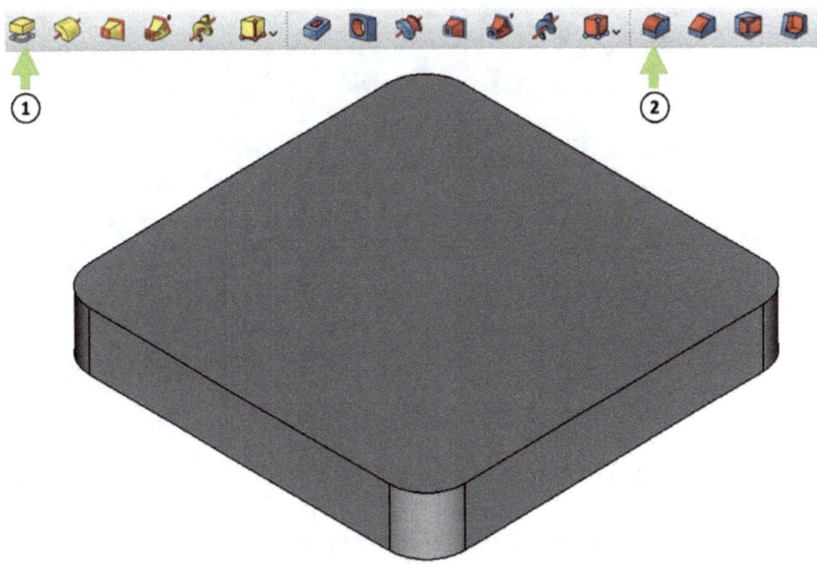

Next, we create a new sketch on the top of the 3D body. Before we draw anything in this sketch, we use the command "Create external geometry". This command can be used to create a reference to an external geometry (e.g., to the outer edges of the 3D body) in the current sketch. We do this by selecting the command ① and clicking on the lines ② - ④ one after the other. This projects the chosen geometries, their end points and any center points into the sketch. These projections are now selectable for the creation of geometry elements or dimensions, for example.

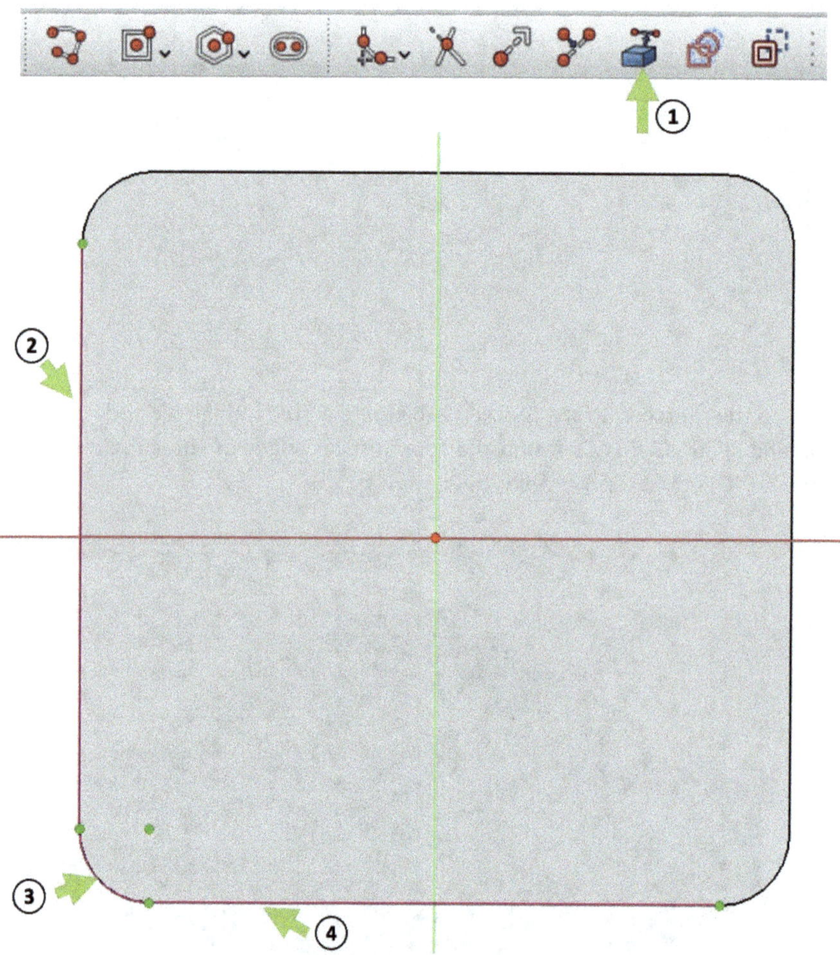

In our case, we then create the following 2D profile by using lines, 3-point arcs, conditions, and dimensions as shown. The 3-point arc ① and the two adjacent lines should lie congruently on the previously projected outer edges of the 3D body.

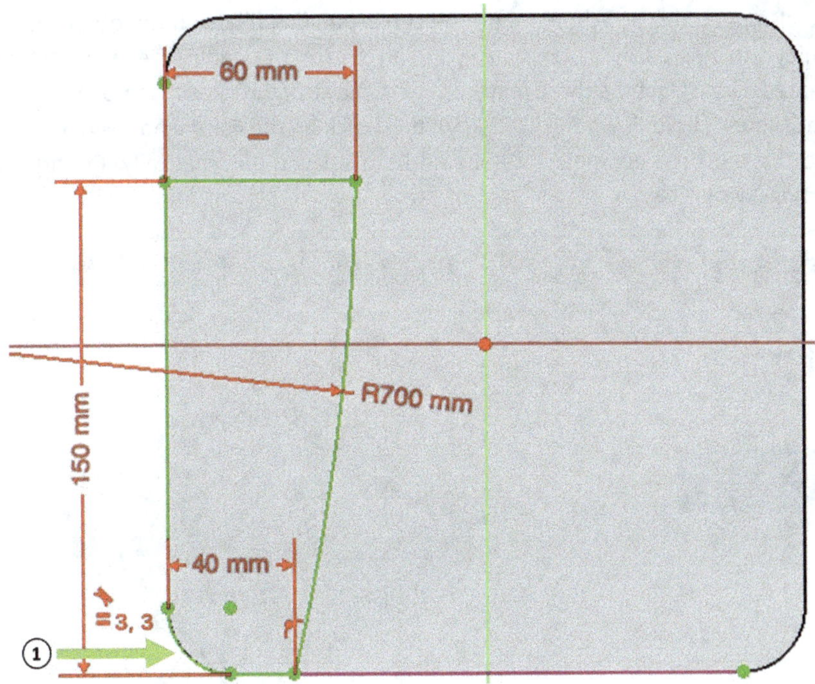

After we have finished the sketch, we can form a 5 mm high attachment with the command "Pad" ① and round the rear edge ③ with a radius of 25 mm as well as the front edge ④ with a radius of 10 mm by using the command "Fillet" ②.

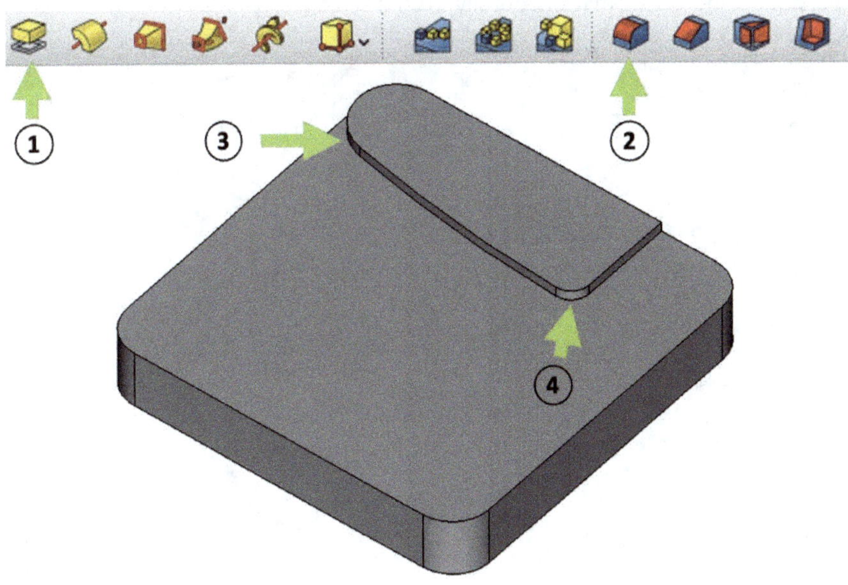

We also need this attachment and the corresponding fillets on the other side. To achieve this as easily as possible, you can mirror the geometries we have just created. First select the three features ① in the structure tree ("Combo View" and tab "Model"; hold down the CTRL key), then click on the command "Mirrored" ② and the mirrored geometry ③ should be displayed automatically. Confirm the settings with "OK".

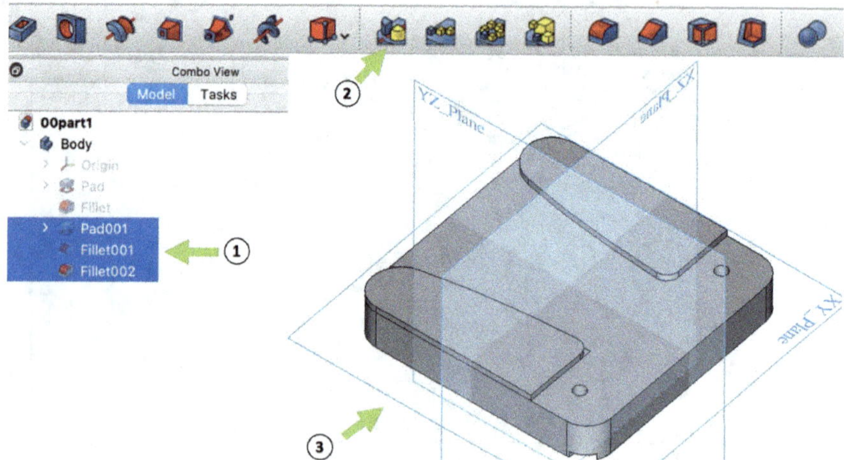

Subsequently, we create another sketch on the top surface of the part. In this sketch, we first project the upper outer edges ② (on both sides) as before ①. Then we create two circles with a diameter of 10 mm, which should sit 18 mm (vertically) and 27.5 mm (horizontally) from the edges.

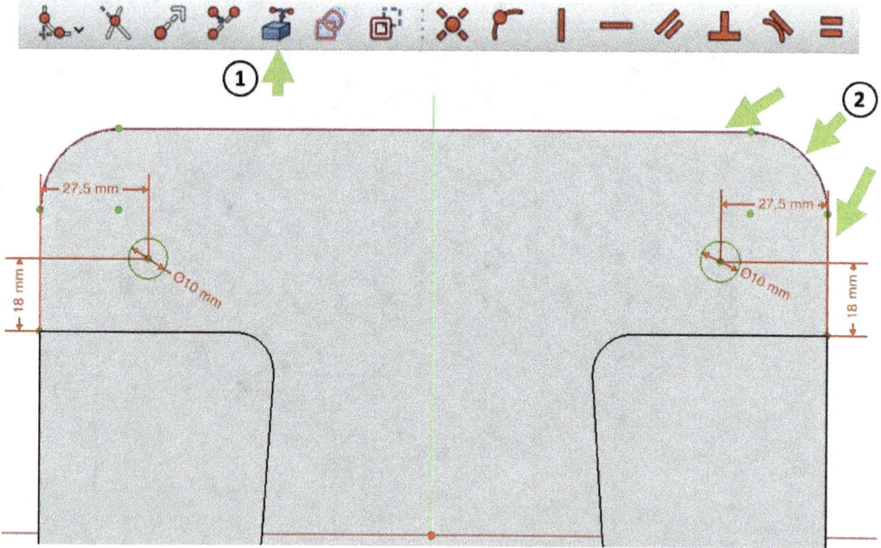

After closing the sketch, we can cut these two circles out of the 3D part using the command "Pocket" ① so that two holes are created through the entire part ("Through all" ② in the settings). Alternatively, we could have used the command "Hole" ③.

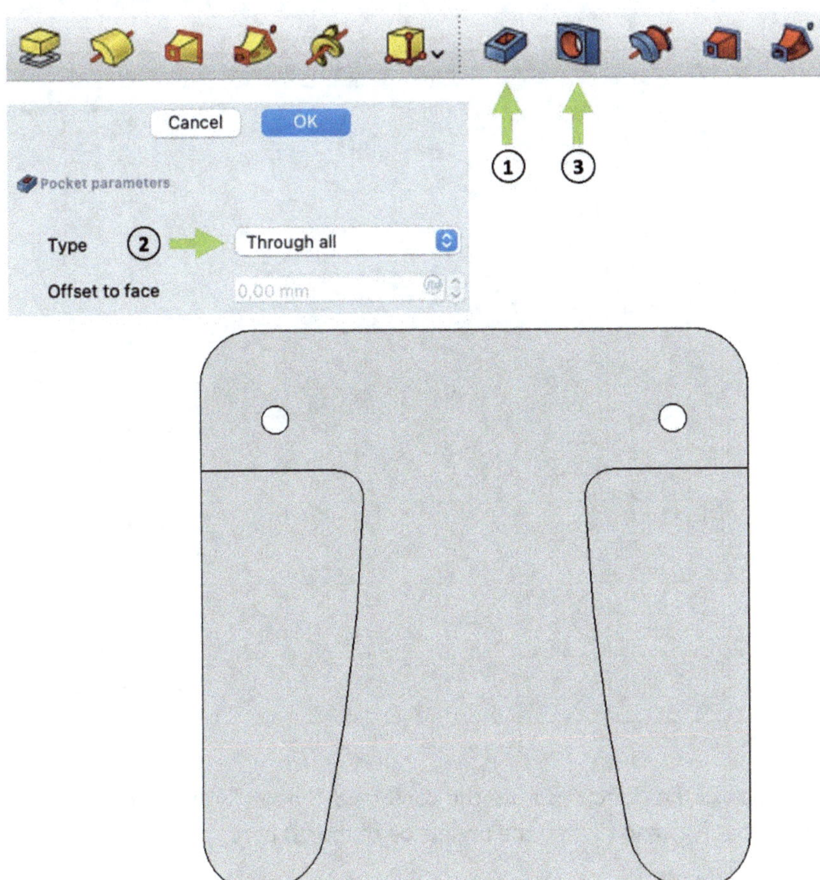

These holes are required for the hole punch in order to transport the punched material into the collecting tray. To ensure that there is enough space for this material, we make a cut-out on the underside of the 3D part in the next step.

Once we have created a sketch on the underside, we first project all the outer edges of the base onto our 2D sketch ①. We then draw a rounded rectangle ("Rounded rectangle") ②, the center of which should lie on the coordinate origin ③. The rounded edges should have a diameter of 15 mm ④. The distances to the outer edges should be 10 mm ⑤ and 20 mm ⑥ at the bottom.

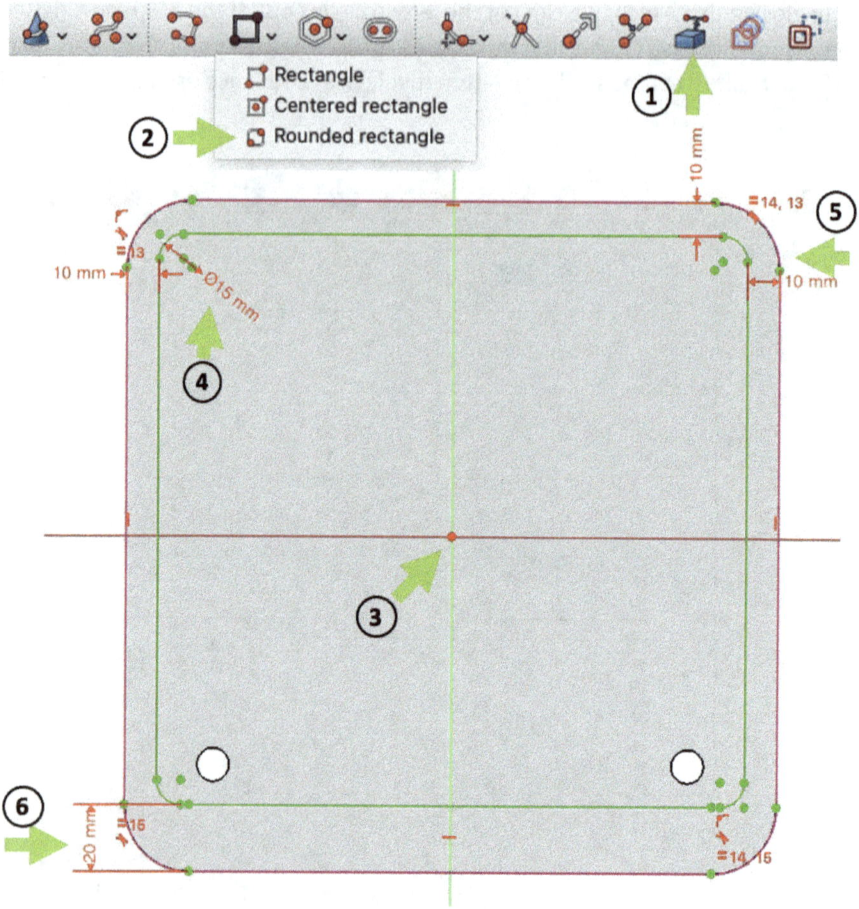

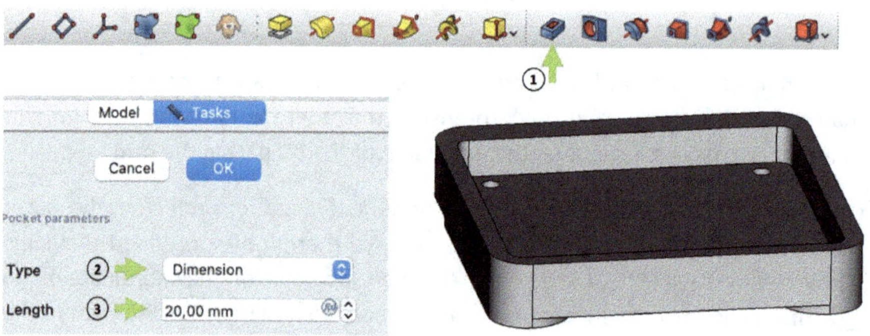

We then close the sketch and use the command "Pocket" ① to create a 20 mm deep cut-out (② and ③) on the underside of the 3D part.

The base of the hole punch is now almost finished. All that's missing is a cut-out on the side where the paper stop rail will sit. To do this, we create a sketch on the right-hand side of the part and draw the following rectangle.

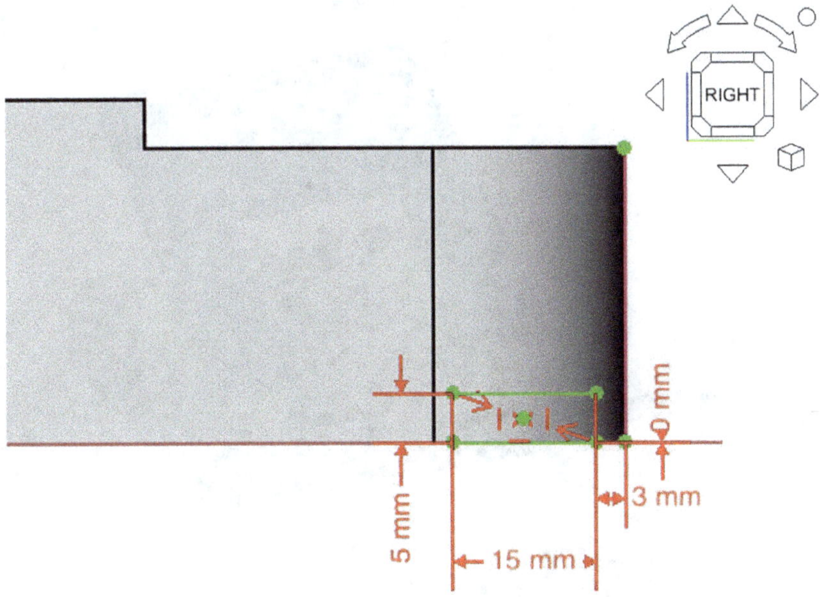

After finishing the sketch, we again use the command "Pocket" ① to create a 150 mm long section (② and ③).

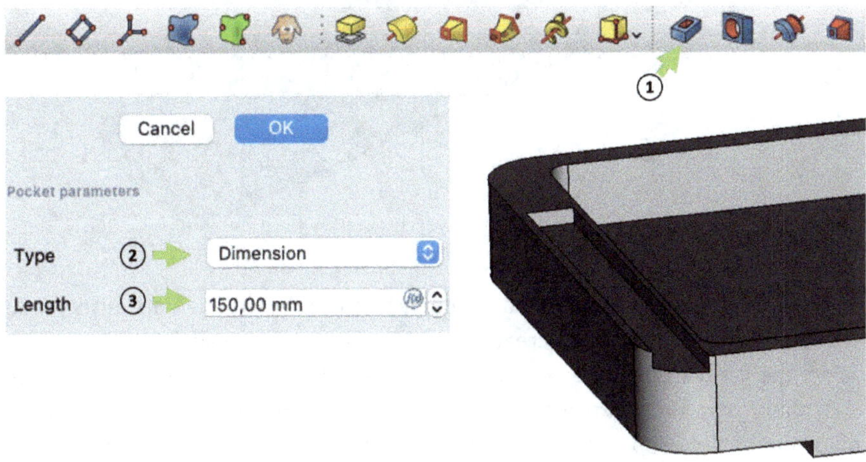

Perfect, the base of the hole punch is done, and we can continue with the second part straight away.

3.2 The brackets for the lever mechanism

To mount the lever mechanism on the base, we need two U-shaped mounting elements.

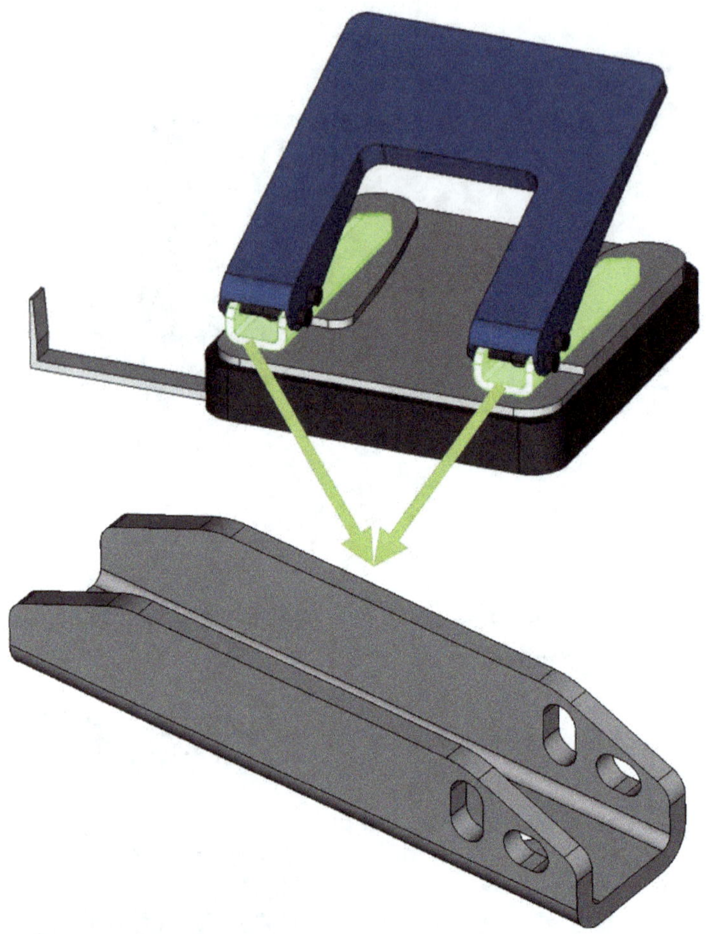

As these brackets are symmetrical, we only need to design one part of them. So, we start in a new document, create a solid and a sketch as usual. In this case, we select the y-z plane for the sketch. We can draw the cross-section of the part on this plane and then extrude it lengthwise. The cross-section consists of the following profile. You can either draw the full profile or just half of it and then mirror it. Otherwise, use the "Constrain Equal" condition for lines of equal length to avoid double dimensions.

Once the profile has been drawn and fully defined, the second step is to round off the bottom four corners. We do this with the command "Constraint-preserving sketch fillet" (①️ and ②️). The difference to "Sketch fillet" is that conditions are retained. After selecting the command, we click on the two lines that form a corner one after the other (example: ③️ and ④️). We do this for all four corners (two inside, two outside) in the lower area.

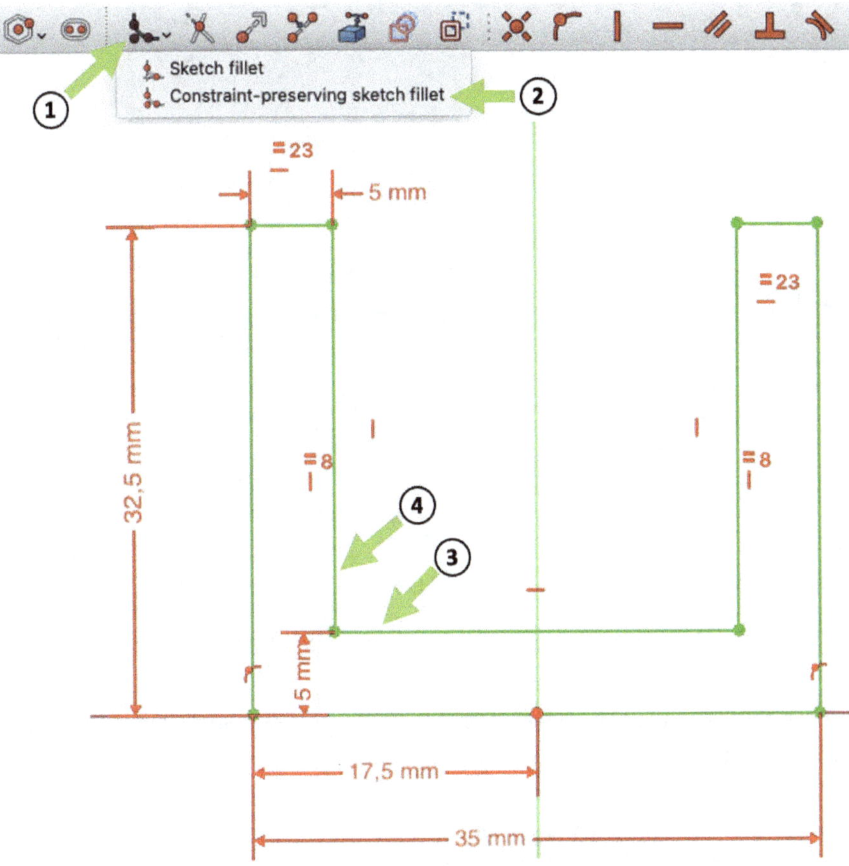

In the last step of the sketch creation process, we need to dimension the fillets we have just created. We do this with the command "Constrain radius" (①️ and ②️). We select a radius of 2.5 mm for the inner fillets and a radius of 7.5 mm ③️ for the outer fillets. However, we only need to do this on one of the two sides, as we can use the condition "Constrain equal" for the other side. You may also have to add one or the other condition so that the profile is fully defined again. It can happen that individual conditions are deleted despite the "**Constrain-preserving** sketch fillet" command.

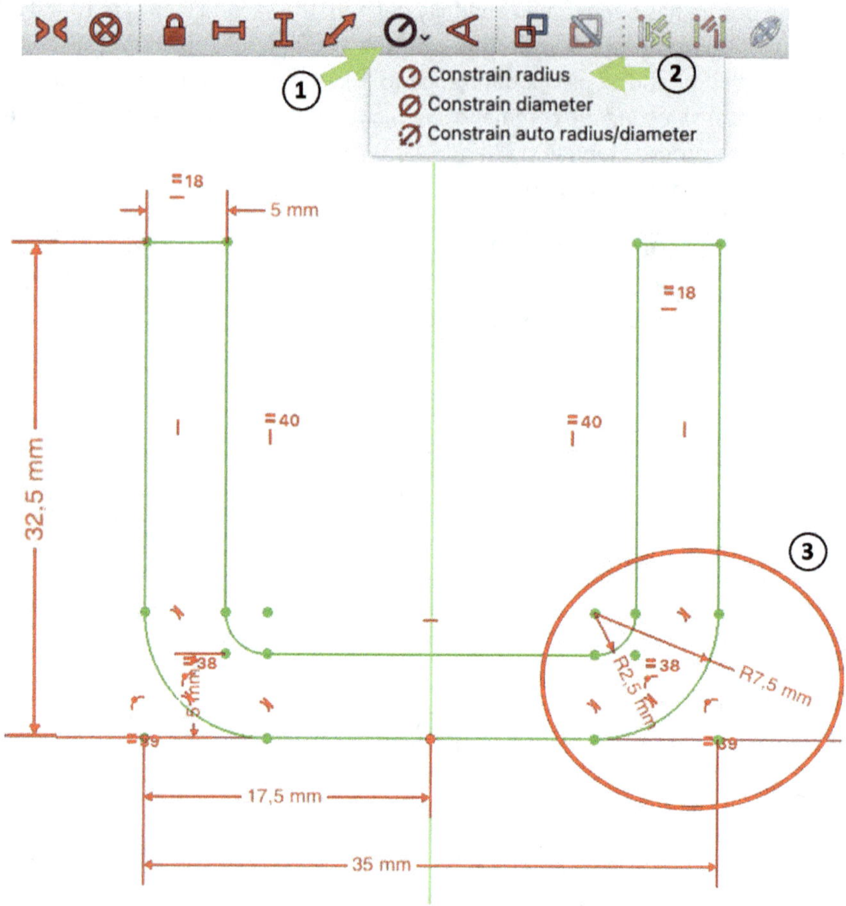

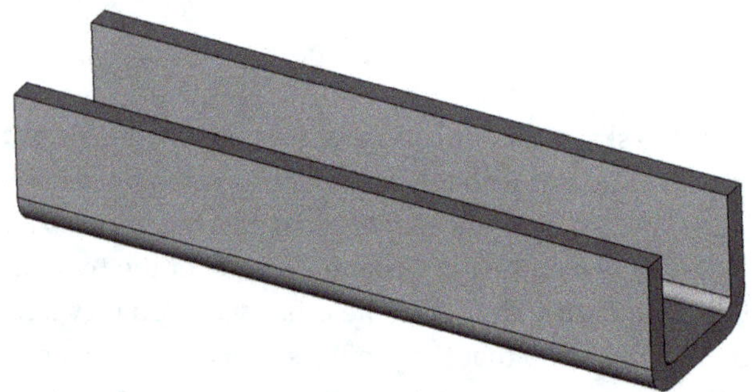

We can then end the sketch with the ESC key and make a linear extrusion of 140 mm. We do this with the well-known command "Pad".

The basic body is done. Now let's get down to refining the 3D part. First of all, we need two slotted holes in which the bolts of the lever mechanism will later sit. To create these slots, we start a new sketch on the front surface of the part. Before sketching, you may need to make a rotation ① and then project the two right edges of the part (③ and ④) into the sketch using the command "Create external geometry" ②. We can use the command "Create slot" ⑤ to easily sketch and dimension the two slotted holes shown.

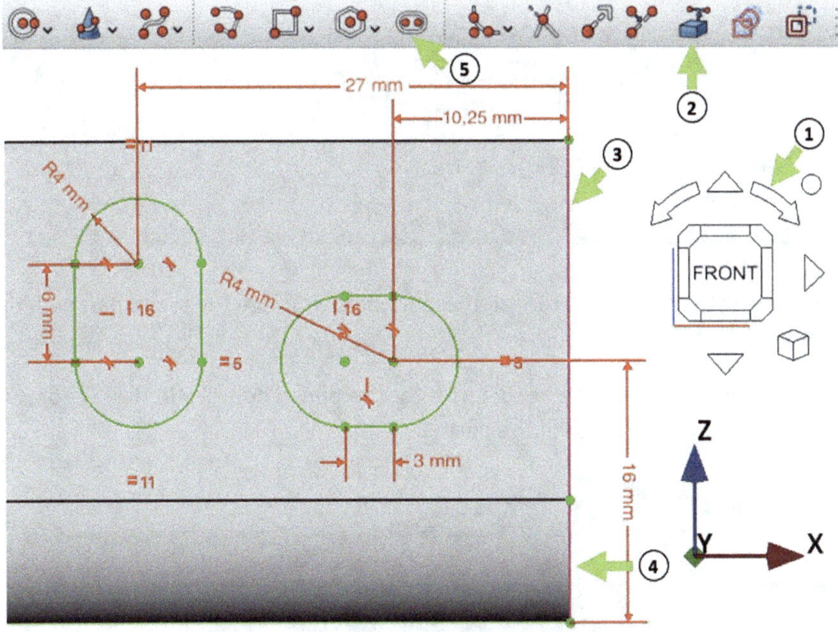

After we have finished the sketch with the ESC key, we can cut the slotted holes out of the part using the familiar command "Pocket". We need the slotted holes on both sides, so we select "Through all" instead of "Dimension" in the command settings.

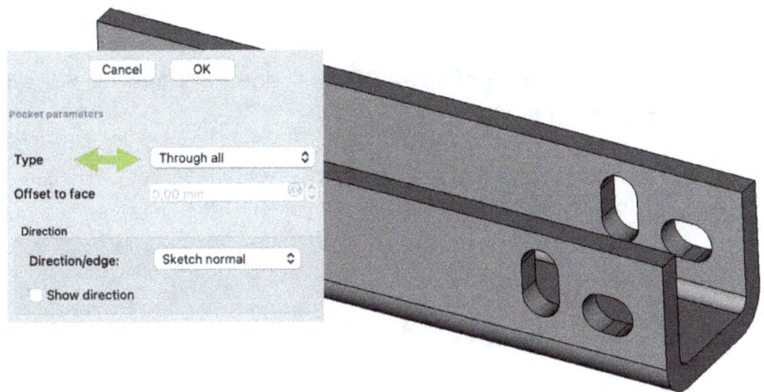

Now we create a cut-out to flatten the two edges in the upper area. To do this, we again need a sketch on the front side of the part. In this sketch, we draw two triangles, each with dimensions of 35 mm and 15 mm. For better dimensioning, first project the top and the two side edges of the 3D part into the sketch (command "Create external geometry").

After closing the sketch, we can cut these triangles through the entire part with the command "Pocket".

In the penultimate step, we need a hole for the punch bolt. To do this, we draw a sketch on the inner surface of the part ①.

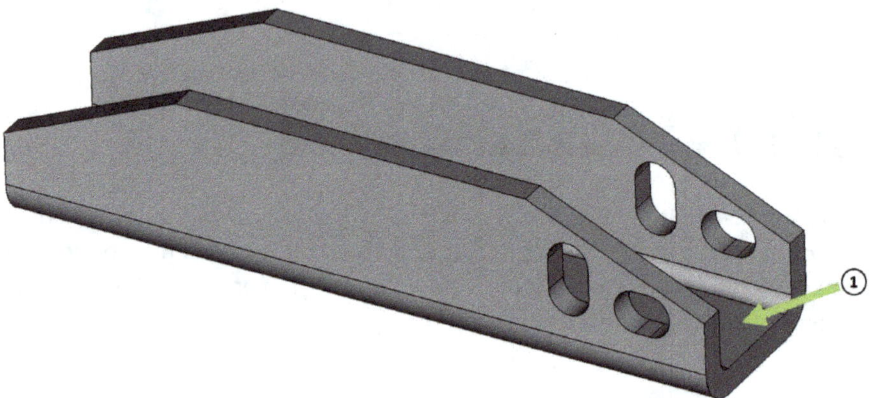

We draw a circle with a diameter of 10 mm on the center axis and set a distance of 27 mm to the right edge of the 3D part.

Next, cut this circle out of the part after exiting the sketch (ESC key). You can simply use the command "Pocket" as usual.

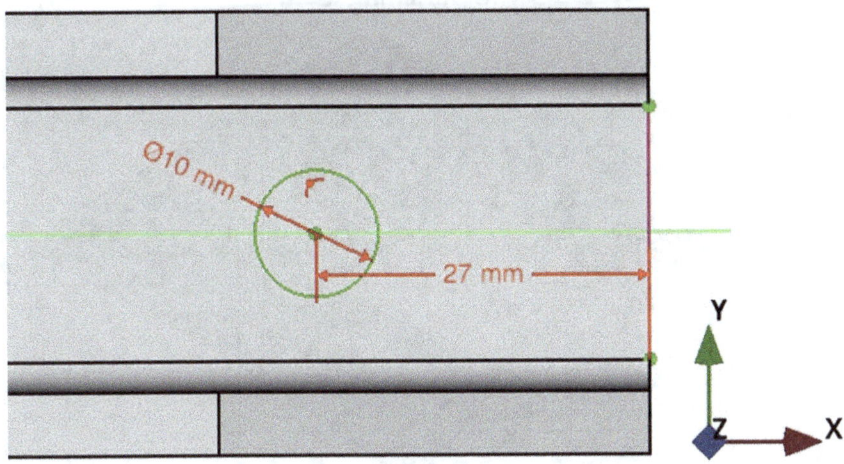

In the last step, we round off the eight upper corners of the part. We do this by using the command "Fillet" twice. First round the corners ① - ④ with a radius of 20 mm and then round the corners ⑤ - ⑧ with a radius of 5 mm.

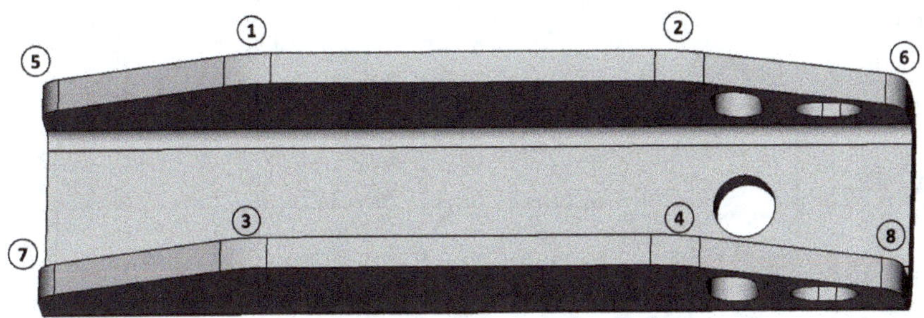

And just like that, we have completed another part of our assembly! But we still need a few components before we can start assembling the hole punch. In the next chapter, we will focus on the punching lever that will be used to operate the hole punch.

3.3 The punching lever

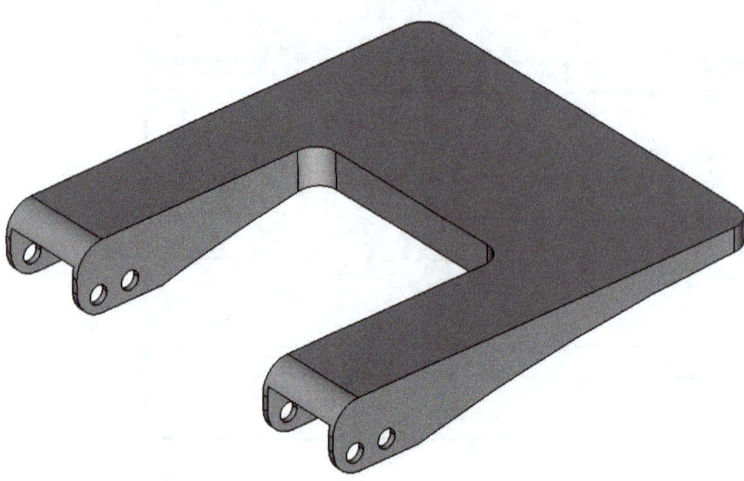

For the 3D part, we start in a new document by creating a body and a sketch on the x-y plane. Create a 200 mm long and 30 mm wide rectangle for the basic body and extrude this rectangle 185 mm with the command "Pad" (close the sketch first).

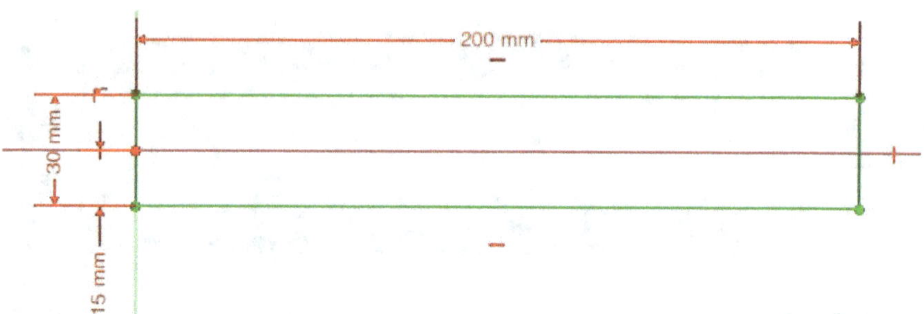

We then create a sketch on the front/upper side surface ① (view "TOP"; see Orbit cube).

Next, we project all edges of the 3D part into the sketch (command "Create external geometry" ①). We then draw two lines (② and ③), starting in the bottom right corner of the part. The line ② should be 10 mm from the top edge and the line ③ 40 mm from the left edge. Finally, we create a 3-point arc between the end points of the lines ("End points and rim point" ④), which gets a radius of 650 mm. The sketch is now fully defined and can be closed.

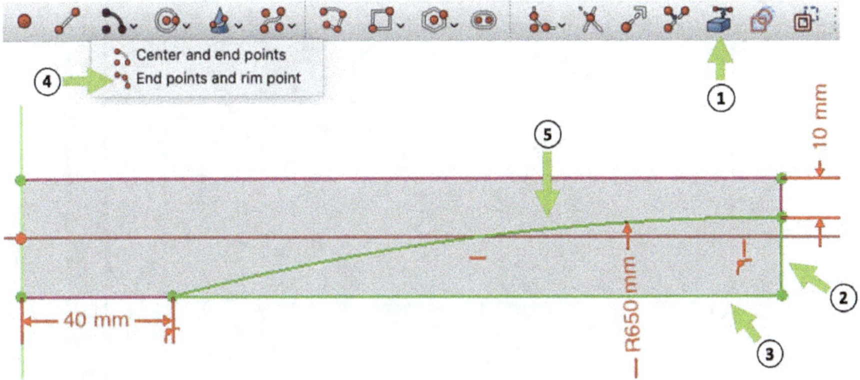

With this sketch and the command "Pocket" let's create a cut-out that goes through the entire part. After rounding the front two edges ① and ② with 12 mm each, the shape shown should result.

We create a new sketch on the upper surface ③ to create the cut-out for the middle part of the punching lever.

Here we sketch a 115 mm wide rectangle with a distance of 85 mm to the rear edge of the 3D part. The rectangle should be centered, so we add a 40 mm gap at the top and bottom. To project the edges, use the command "Create external geometry" once more.

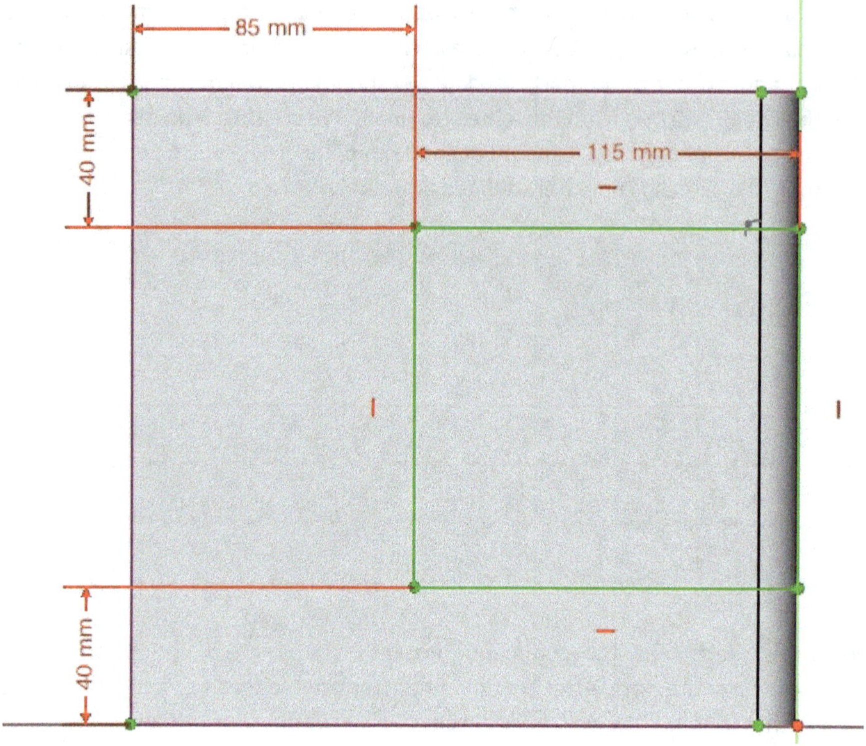

Exit the sketch (ESC key) and use the command "Pocket" to cut through the entire part. As we now look at the part from below, we can round the edges ① - ④ with the command "Fillet" and a radius of 10 mm. We then create a sketch on the surface ⑤.

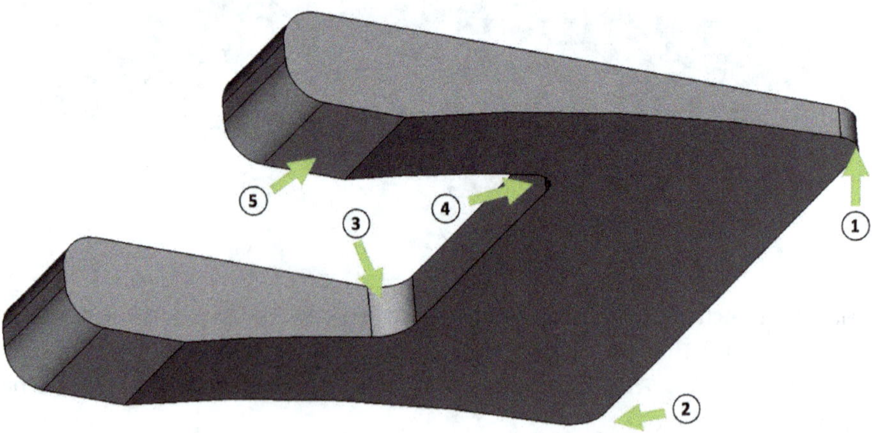

In this sketch, we draw two 200 mm long and 35 mm wide rectangles, which we place at a distance of 30 mm horizontally and 2.5 mm vertically on the two arms of the lever. To be able to set the dimensions, you must first project the edges of the 3D part into the sketch (command "Create external geometry").

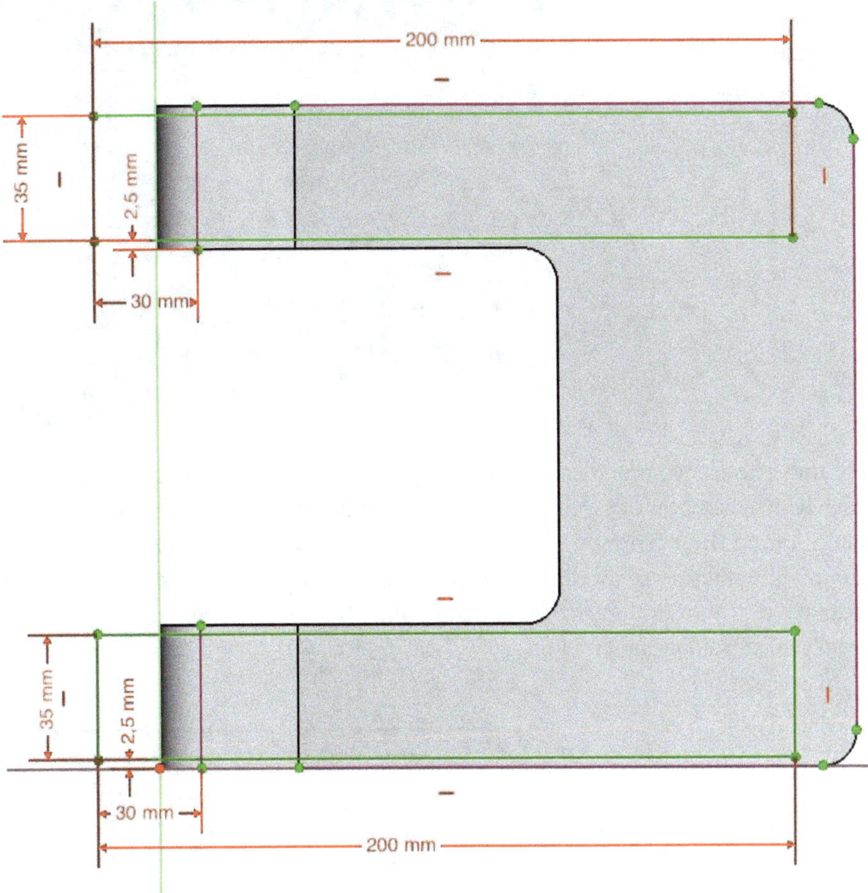

After we have finished the sketch with the ESC key, we can perform the cut-out using the command "Pocket". We need a length of 22 mm, which we can enter in the settings of the command ①. In this case, the option "Type" must be set to "Dimension".

We now need to create another sketch for this part (on surface ②), as we still require two through holes for the bolts.

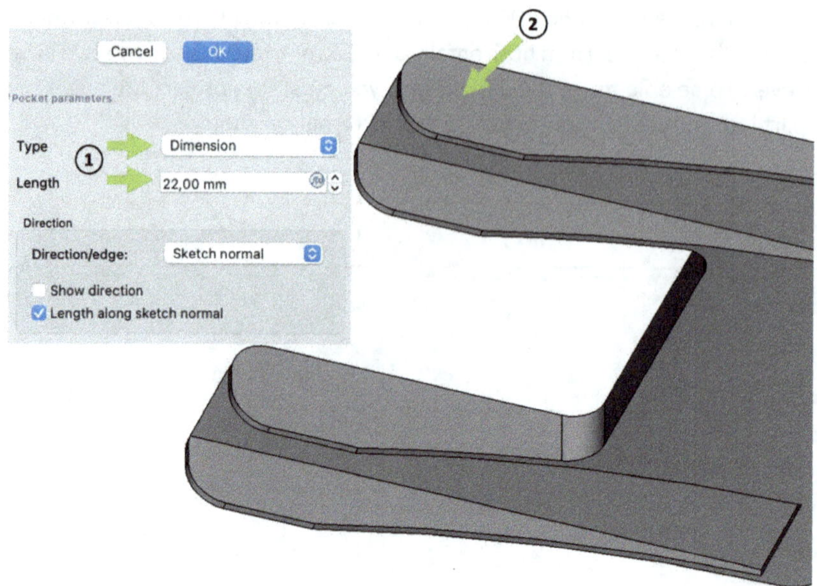

For these holes, we require the following geometry, which consists of two circles, each with a diameter of 8 mm. The centers of the two circles should have a distance of 15 mm in the horizontal direction and be at the same height vertically. For a complete definition of the sketch, the left-hand circle center point is also given a distance of 7 mm in the vertical direction and 10 mm in the horizontal direction from the coordinate origin ①.

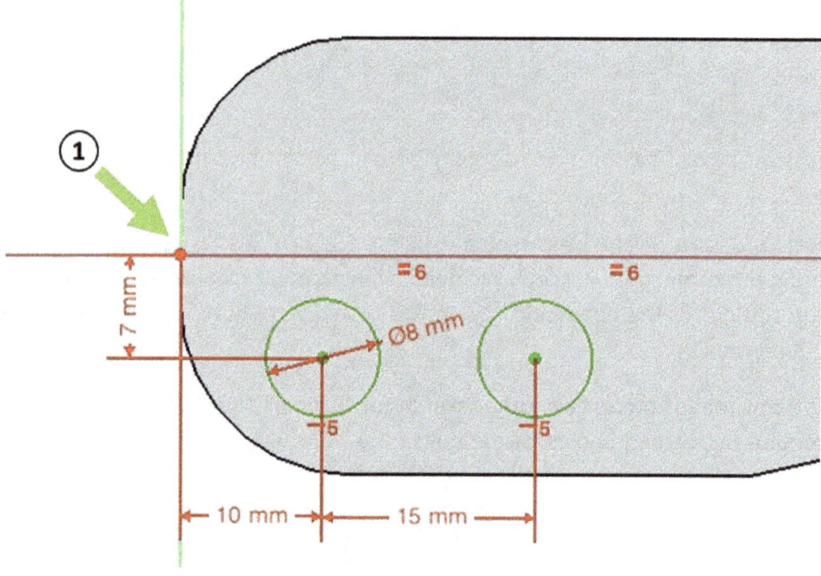

When you have finished drawing, all you have to do is to create the holes — as usual — using the command "Pocket" with the setting "Through all". This will create the holes on the other side of the 3D part as well.

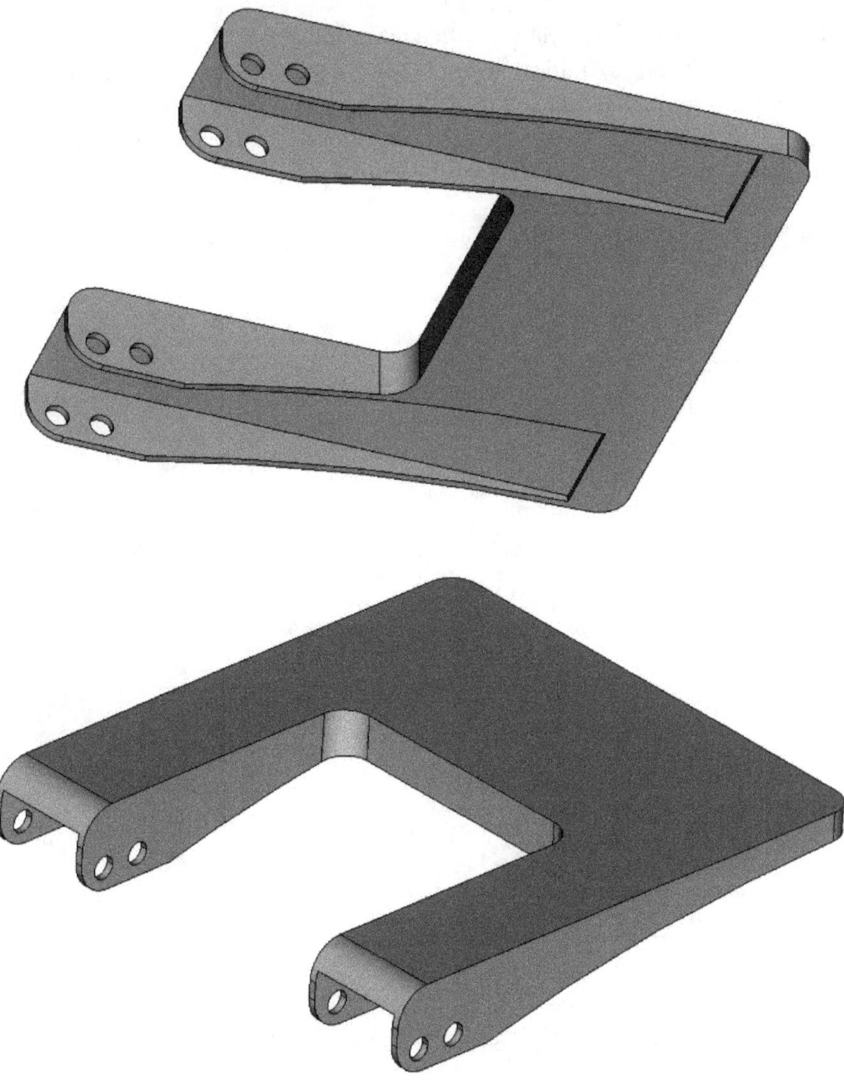

Perfectly done! This part was a little more complex to construct, but in the next chapter the simplest components of the hole punch follow. We'll take care of the bolts, the spring, and the punch bolt. When we're done with that, we'll assemble the hole punch — so, look forward to another successfully completed project! Let's go!

3.4 Bolt, coil spring and punch bolt

We need two bolts, coil springs and punch bolts for the hole punch. These are geometrically identical, and we therefore only need to design one part each.

The bolt is a 50 mm long cylinder with a diameter of 8 mm, at both ends of which there is a 5 mm long cylindrical thickening with a diameter of 10 mm. We construct the part as a rotational part and therefore first draw a quarter of the cross-section on the x-z plane ①. End the sketch with the ESC key and then click on the command "Revolution" ② to create the 3D part ③.

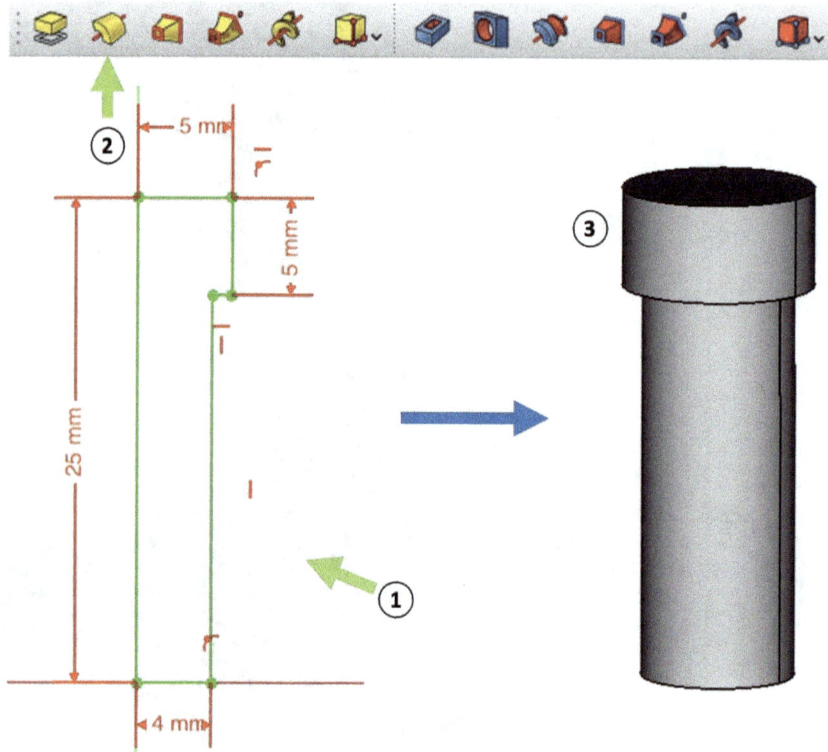

To obtain the full bolt, we need to mirror this half. We do this by clicking on "Mirrored" ① and selecting "Base XY plane" ② in the "Plane" option in the settings.

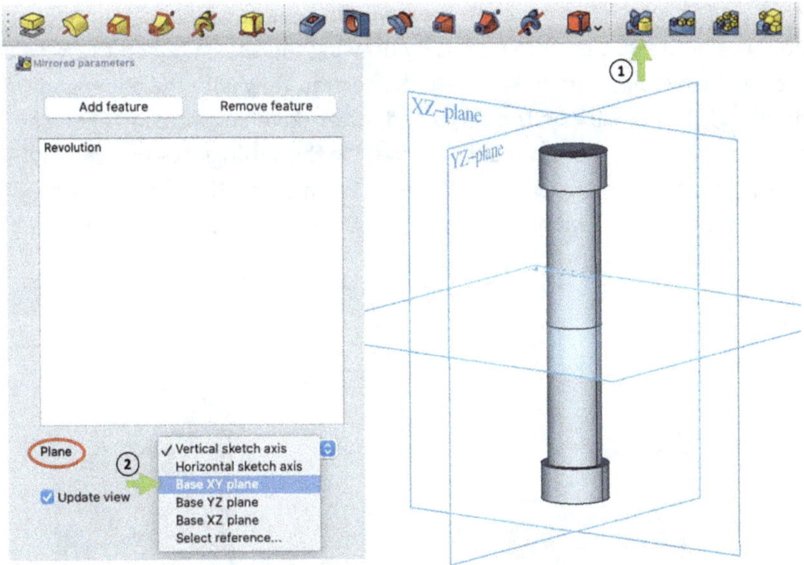

The bolt is done. Let's continue with the punch bolt. The basic body is a 26 mm long cylindrical body with a diameter of 10 mm. The underside of the cylinder is given a conical indentation so that the paper can be punched out. To do this, we draw the sketch ① on the x-y plane and obtain the part ③ after closing the sketch and using the command "Revolution" ②.

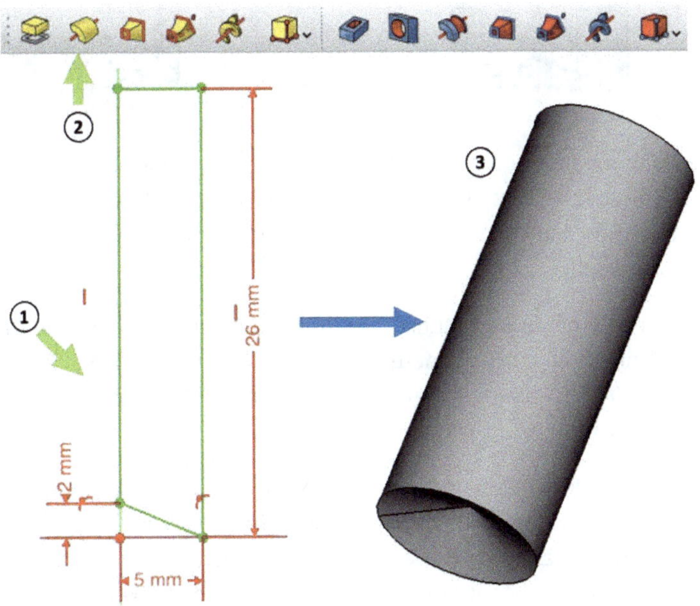

So that we can attach the punch to the bolt, we need to add an 8 mm hole in the upper area of the 3D part. We do this by drawing the sketch ② on the x-y plane. The easiest way to it, is to first activate the section view with the command "View section" ①. Simply click on the command. After sketching, press the ESC key and cut out the hole with the command "Pocket". In the settings, we must select "Two dimensions" and enter two dimensions (e.g., 5 mm each). The setting "Through all" does not work here, as we have sketched on the center plane of the part.

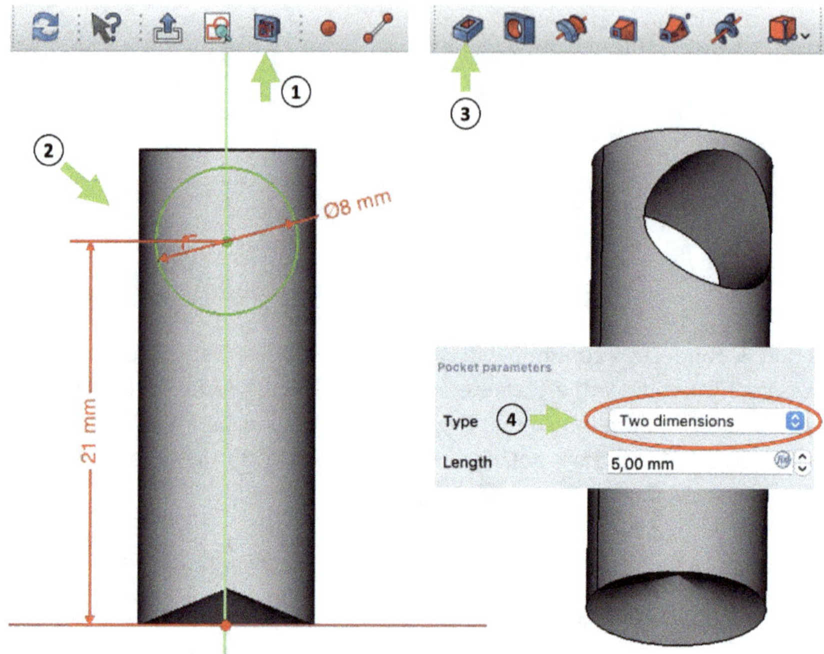

The only thing missing now is the coil spring. The procedure for creating such a spring was already described in the previous book as a first design project. You are welcome to try out the coil spring on your own first. You can also think about the dimensions yourself.

Otherwise, simply continue reading. The punch bolt must be located inside the spring, i.e., it needs an inner diameter of at least 10 mm. We can set the thickness of the spring wire to 1.5 mm, for example. The length of the spring in the relaxed state must be 17 mm.

After we have created a body in a new document, we draw the 2D geometry that we will later extrude along a helix (spiral path). To do this, we create a sketch on the x-z plane. Here we simply draw a circle with a diameter of 1.5 mm (thickness of the spring wire) at a distance of 5.75 mm (to achieve the 10 mm inner diameter, we need half the diameter and 1.5/2 = 0.75 mm as an allowance for clearance between bolt and spring) to the coordinate origin.

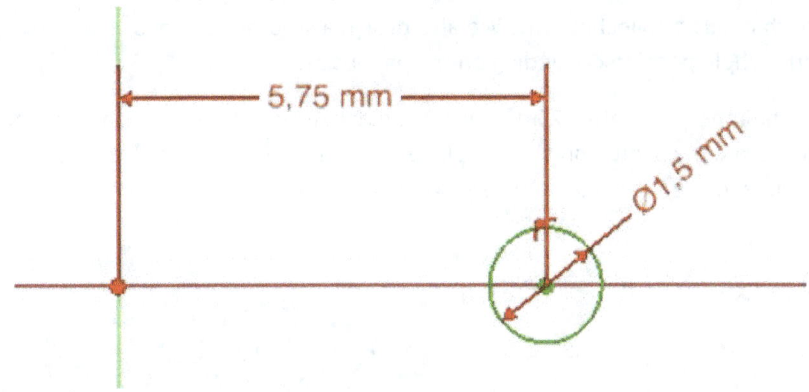

Once we have completed this sketch, we can use the command "Additive helix" ①. We can define the helical spring using three parameters. By default, after clicking on the command, the sketch profile is rotated along the vertical sketch axis and the option "Pitch-Height-Angle" is selected ②. For the parameter "Height" ③ we enter 12 mm and for the parameter "Pitch" ④ we select e.g., 3 mm. Click on "OK" to create the 3D model.

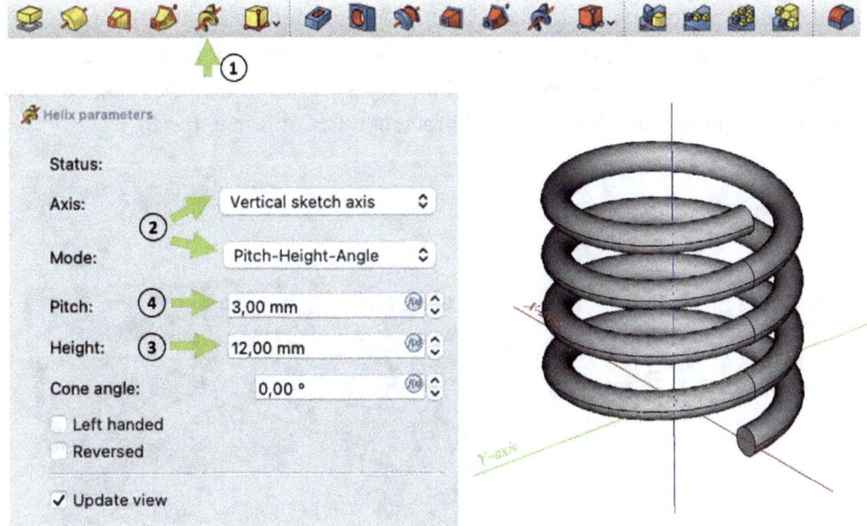

Perfect, only two components are missing before we can start assembling the hole punch. Stay tuned, we will soon have successfully completed this project too!

3.5 Collecting tray and paper stop rail

To ensure that the punched material does not leave any waste behind, our hole punch needs a collecting tray. We also design a stop rail that allows us to set the correct hole position depending on the paper size.

For the tray, we create a 25 mm high cuboid with a square base area of 200 × 200 mm. Draw the sketch on the x-y plane (rectangle with center at the coordinate origin) and use the "Pad" command as usual.

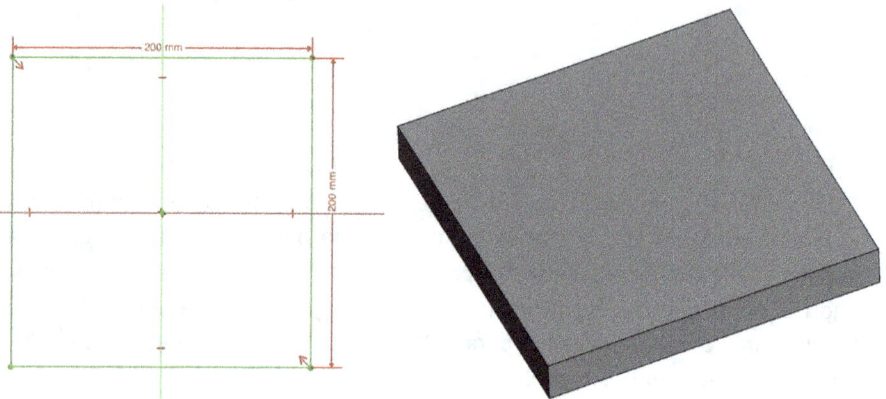

We then use the command "Thickness" ① to hollow out the 3D body evenly. After activating the command, we click on the top ② of the body and enter the value 5 mm for the option "Thickness" ③, which gives us a 5 mm thick wall. We also change the option "Join Type" ④ to "Intersection". Confirm with "OK".

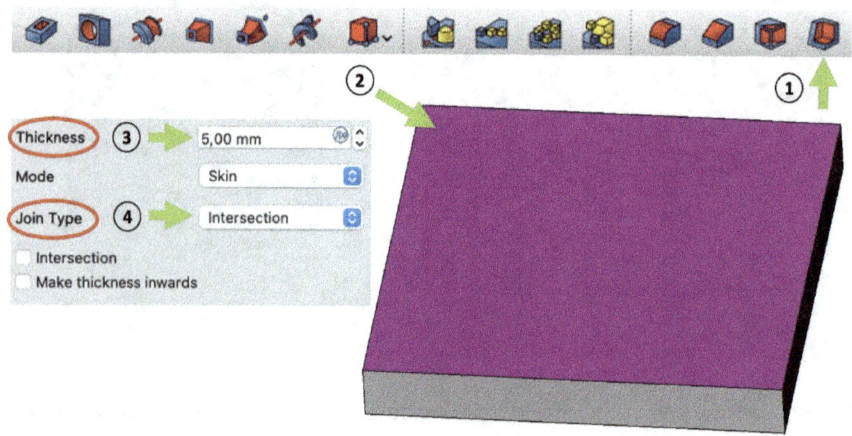

We then round the four inner corners (1) and the four outer corners (2) with a radius of 20 mm (command "Fillet"). The only thing missing is a cut-out for the paper stop rail. For this, we create a sketch on the left side surface (3).

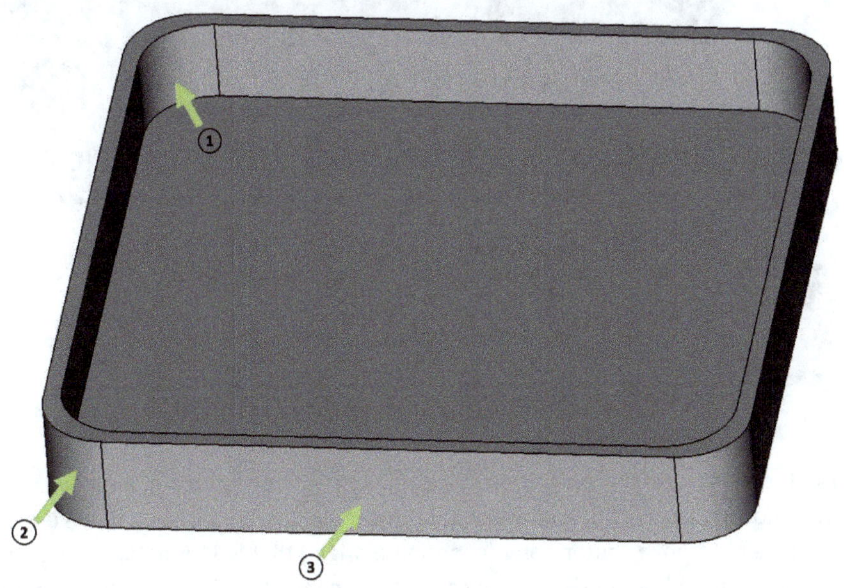

In this sketch, we draw a rectangle 15 mm long and 5 mm wide, which should have a distance of 82 mm from the coordinate origin (1). The rectangle should be to the right of the coordinate origin, and the bottom line of the rectangle should be congruent with the y-axis.

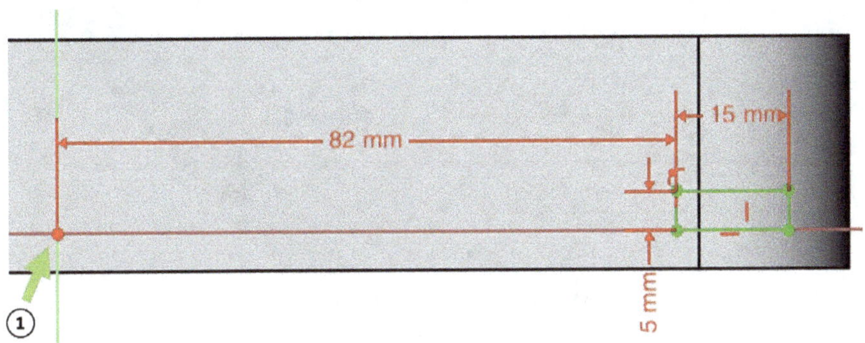

We can then finish the sketch and cut the profile out of the collecting tray using the command "Pocket". The cut-out should go through the entire side wall.

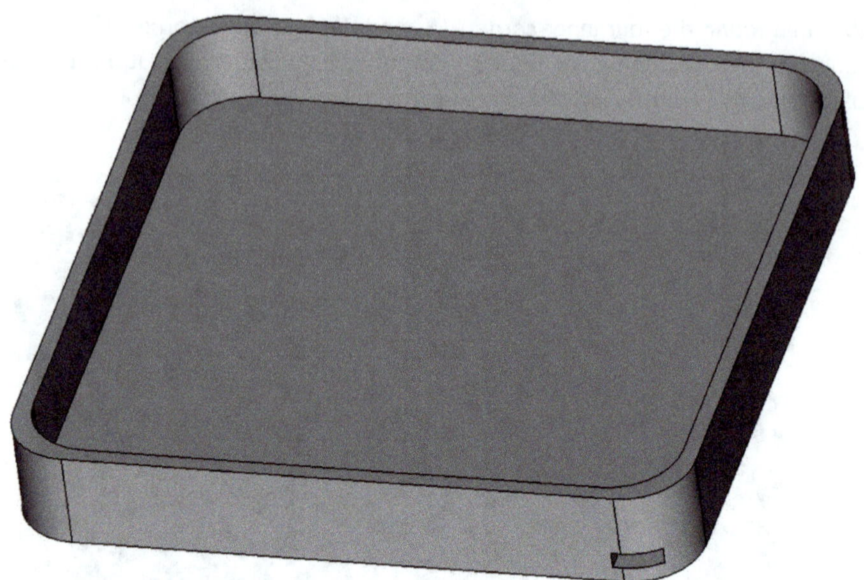

Save the part under any name before getting to the paper stop rail. First, we need a new document, a solid, and a sketch on the x-y plane. In this sketch, we draw a rectangle ① from a center point (coordinate origin) that is 15 mm long and 5 mm wide. After finishing the sketch, we can use the command "Pad" to transform the rectangle into a 3D part ②. For further processing, we create a sketch on one of the two side surfaces ③ (here: left side surface).

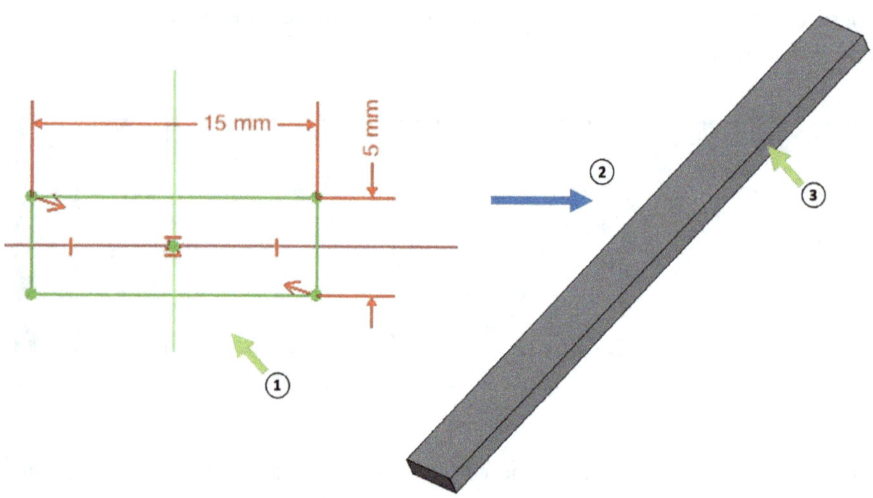

In this sketch, we draw the following profile for the rail stop connection in the upper section of the 3D part.

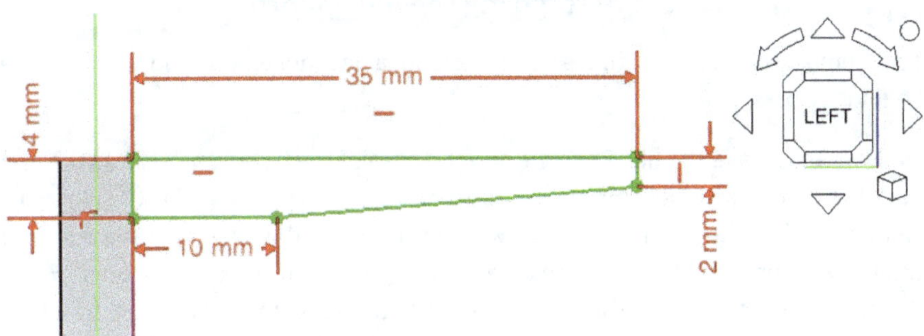

To finish the part, we must now select the command "Pad" after finishing the sketch in order to extrude the sketch with a length of 15 mm. Depending on the orientation, you may need to check or uncheck the option "Reversed".

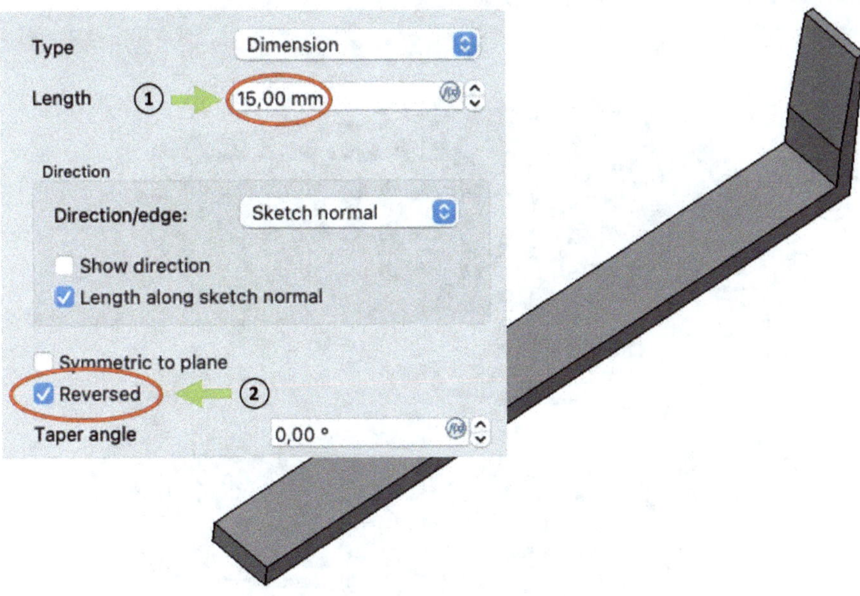

The paper stop rail was the last individual part in this project. In the following chapter, we can therefore take care of assembling all components of the hole punch.

3.6 Assembling the hole punch

To assemble the hole punch, we need to create a new document and switch to the "A2plus" workspace.

First, we must save the file. Then we add the base of the hole punch ② as the first part. We do this with the command "Add a part from an external file" ①. This part will be fixed in the assembly. Next, we add the collecting tray ③. We rotate it using the command "Move the selected part" ④ so that the two cut-outs ⑤ for the paper stop rail are on the correct side. We then create the first constraint for the two parts by clicking on the plane ⑥ and the plane ⑦ one after the other while holding down the CTRL key and selecting the command "Add PlaneCoincident Constraint" ⑧.

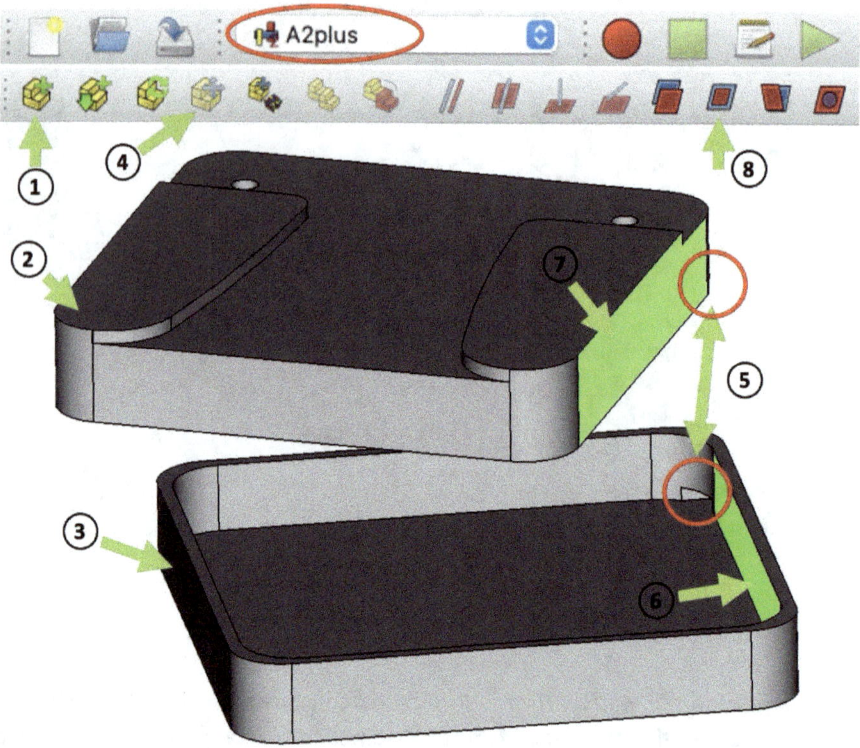

Furthermore, we create such a constraint ③ between the front surface ① of the base and the inner front surface ② of the collecting tray.

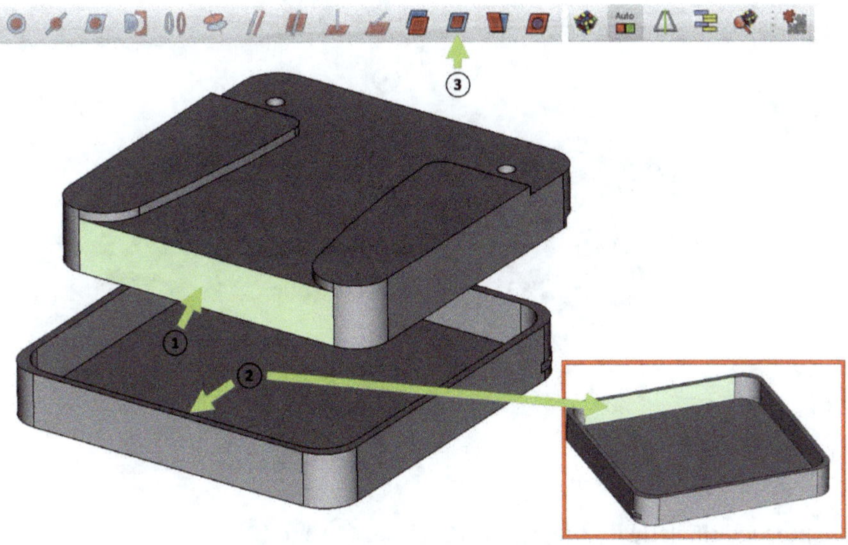

After connecting the inner base surface ① and the lower edge surface ②, the tray is fully mounted to the base.

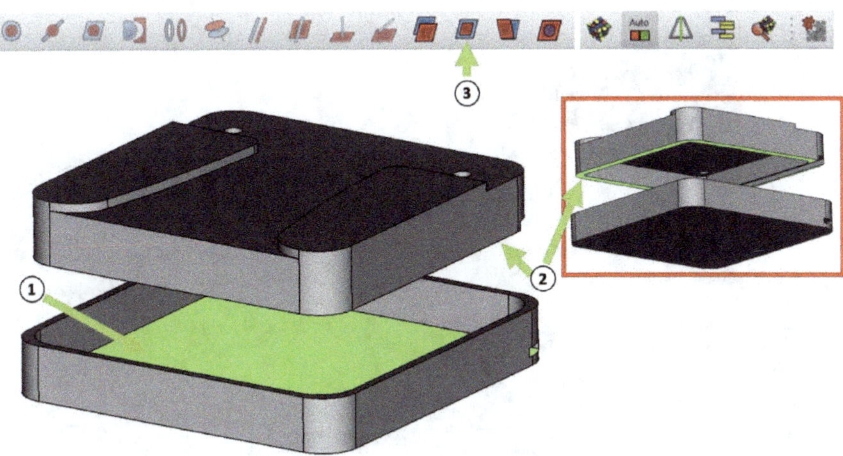

Subsequently, we can insert and mount the next part, e.g., the paper stop rail. First, we rotate the rail so that it is positioned approximately as shown. We then click on the surfaces ① and ② and select the command "Add PlaneCoincident Constraint" ③.

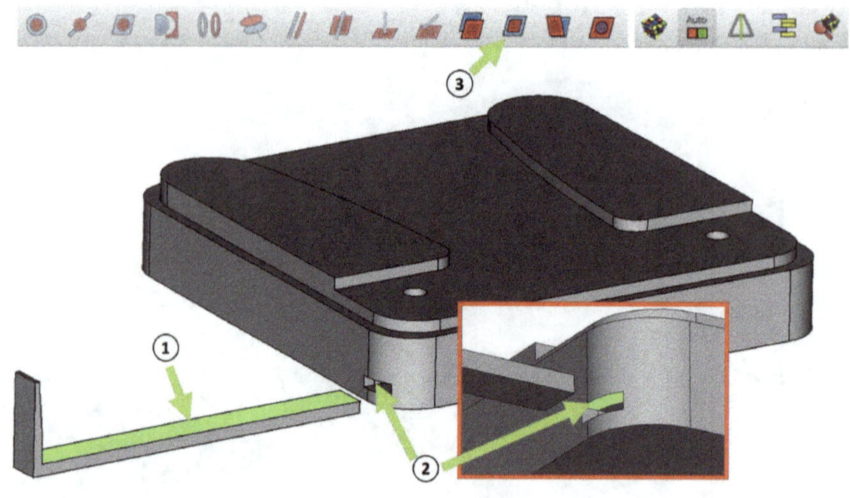

We proceed similarly for the two rear surfaces ① and ②.

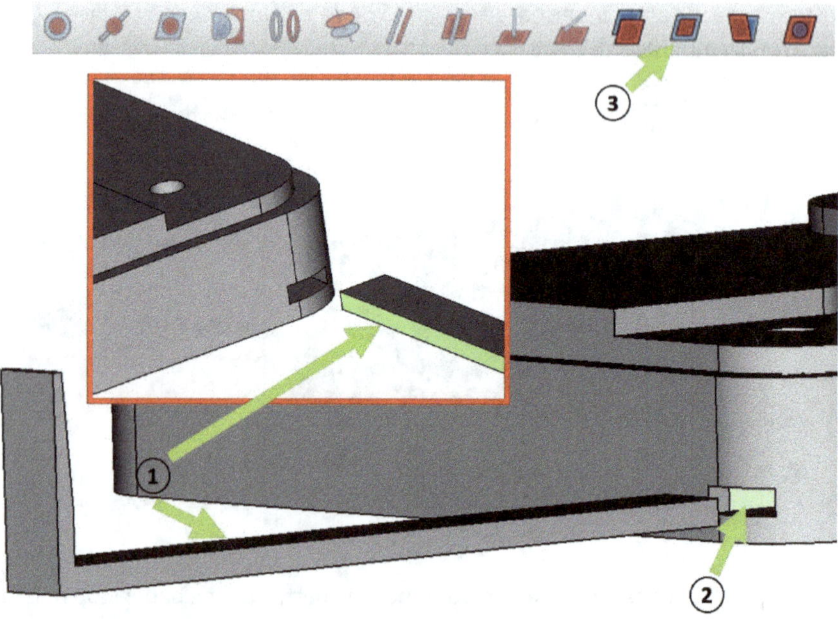

For the complete installation of the stop rail, we need one more link using the command "Add PlaneCoincident Constraint" (① - ③). Here, however, we need to specify an offset (e.g., 100 mm) in the command settings so that the rail protrudes a bit.

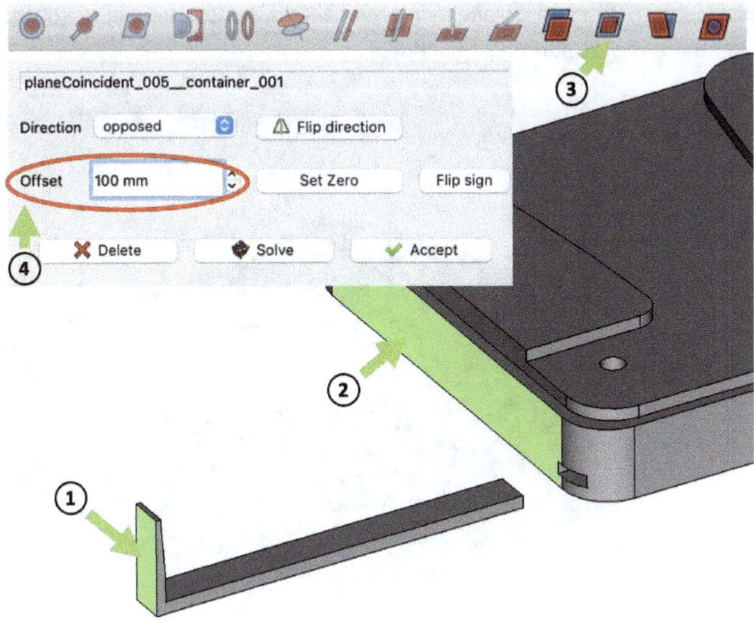

As the next part, we insert one of the two brackets for the lever mechanism into the assembly and first align the two holes ① and ② with the command "Add AxisCoincident constrain" ③.

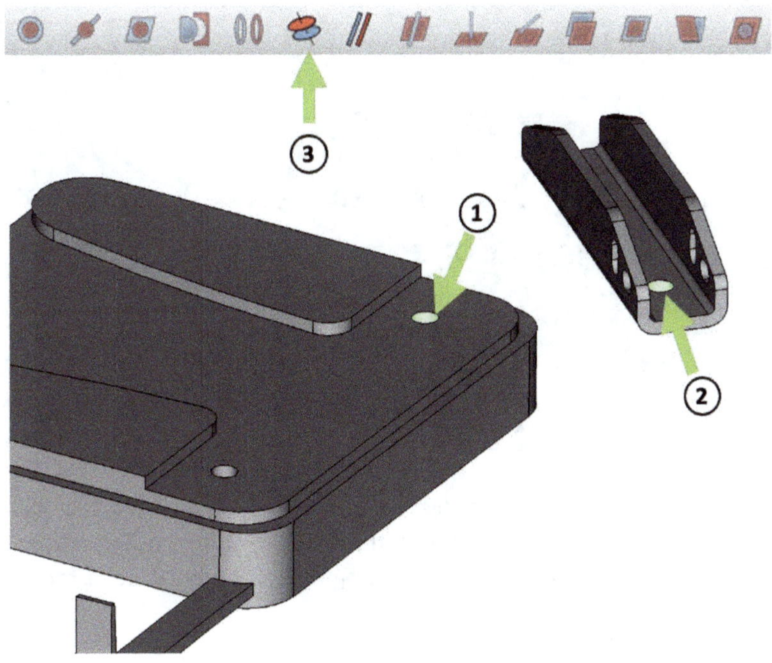

We then need to create a parallel constraint ("Add PlanesParallel Constraint") ③ between the two side surfaces ① and ②.

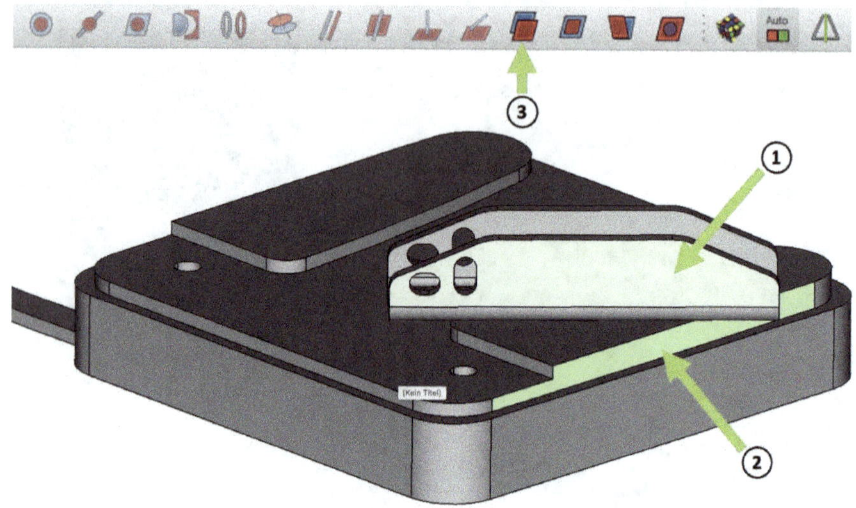

Lastly, we mount the bracket on the base (① - ③).

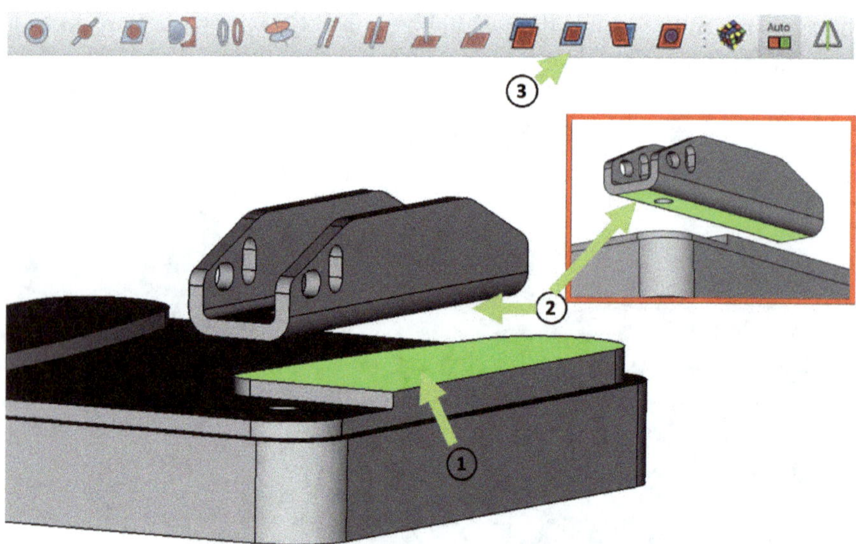

A bracket for the other side is missing. We can easily achieve this with the command "Duplicate" ② (first select the bracket ① in the structure tree). Afterwards, link this second bracket ③ using the same procedure as for the first bracket.

Next, let's insert and mount the punching lever. Rotate the lever so that it is positioned approximately as shown and then align the holes ① and ② with the command "Add AxisCoincident Constraint" ③.

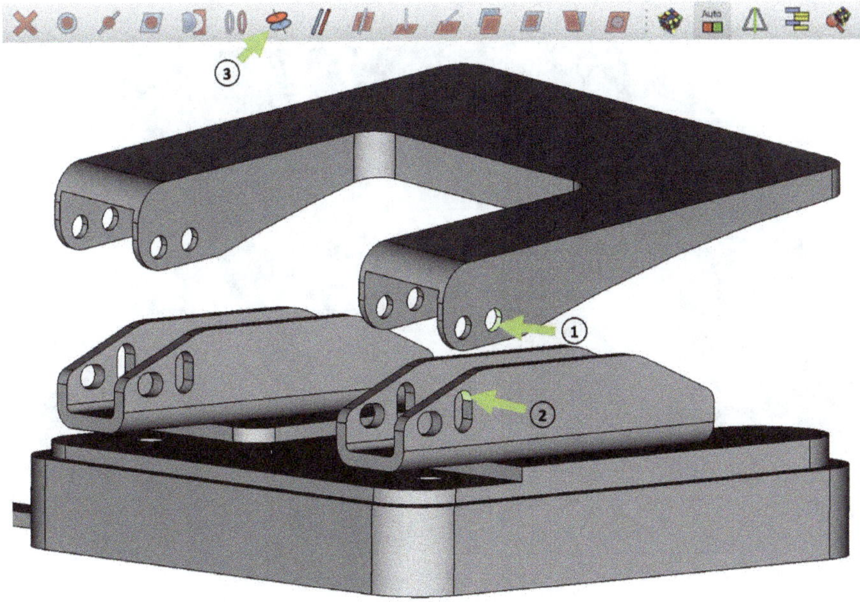

We also set such a constraint for the other two holes on the same side.

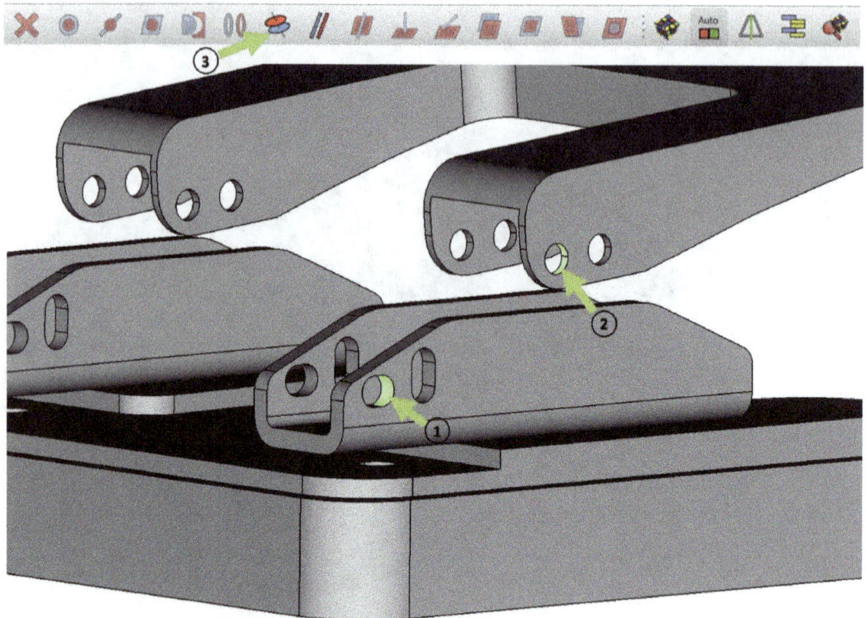

The only thing missing for complete fixation of the part is a link between the inner side surface of the lever and the outer side surface of the bracket (① - ③).

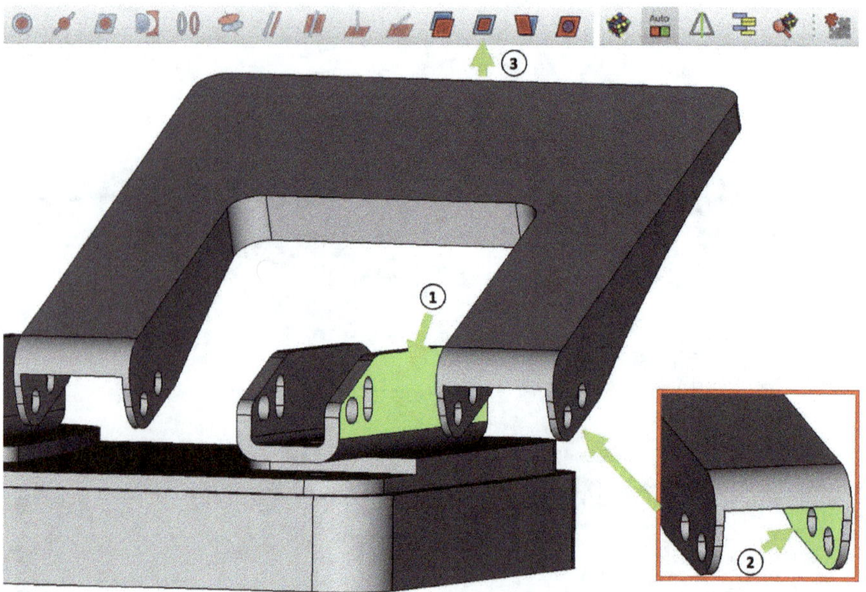

The next part that we insert into our assembly is one of the bolts. In the first step, we can make it ① concentric to the hole ② of the punching lever using the command "Add AxisCoincident Constraint" ③.

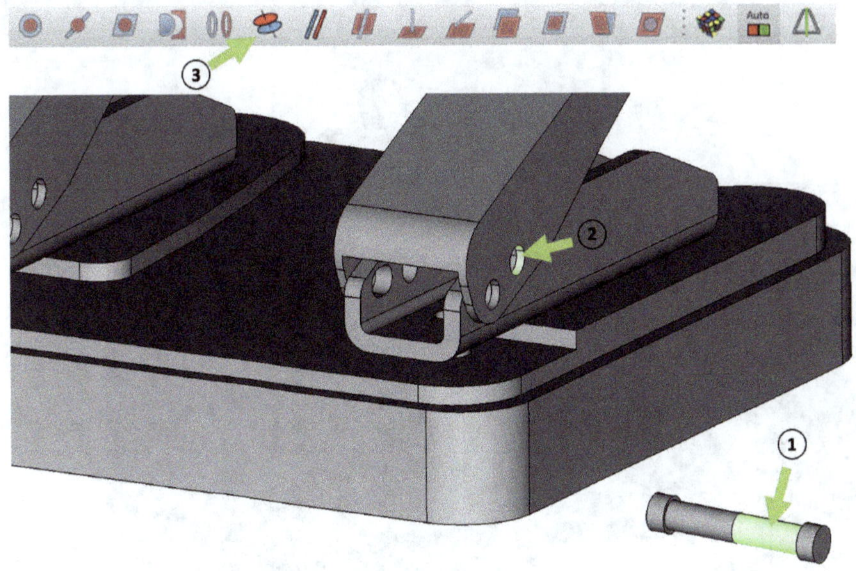

In the second step, we can link the inner surface ① of the bolt with the outer side surface ② of the lever, as shown (① - ③).

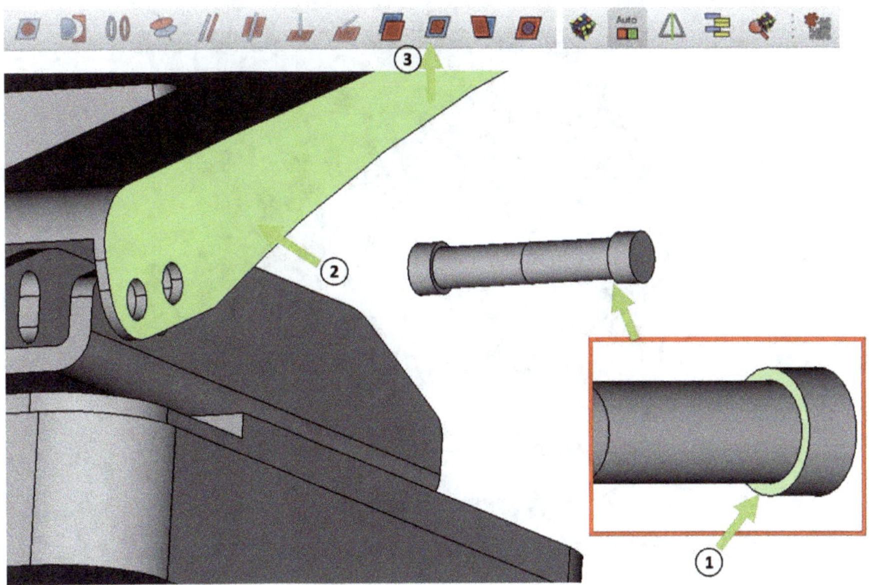

We then add one of the two punching bolts to our assembly and link it with two concentric constraints (command: "Add AxisCoincident Constraint").

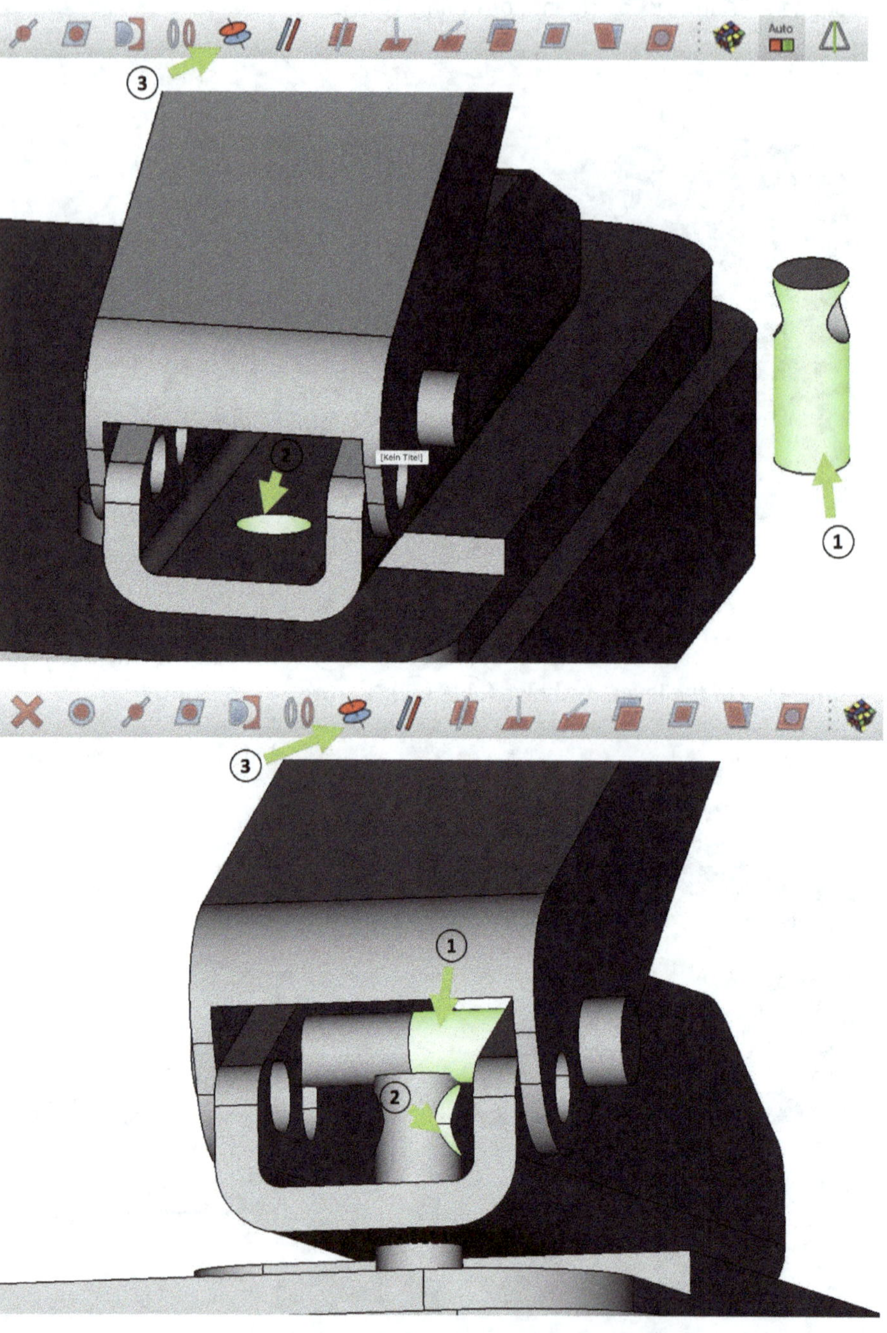

Next comes one of the two coil springs. Unfortunately, we cannot connect these, but we can move them to the correct position. This requires a little dexterity and patience — depending on how perfectly it should sit. We use the command "Move the selected part" ① for this purpose. It allows rotations and movements to be made by dragging the arrows and balls ②.

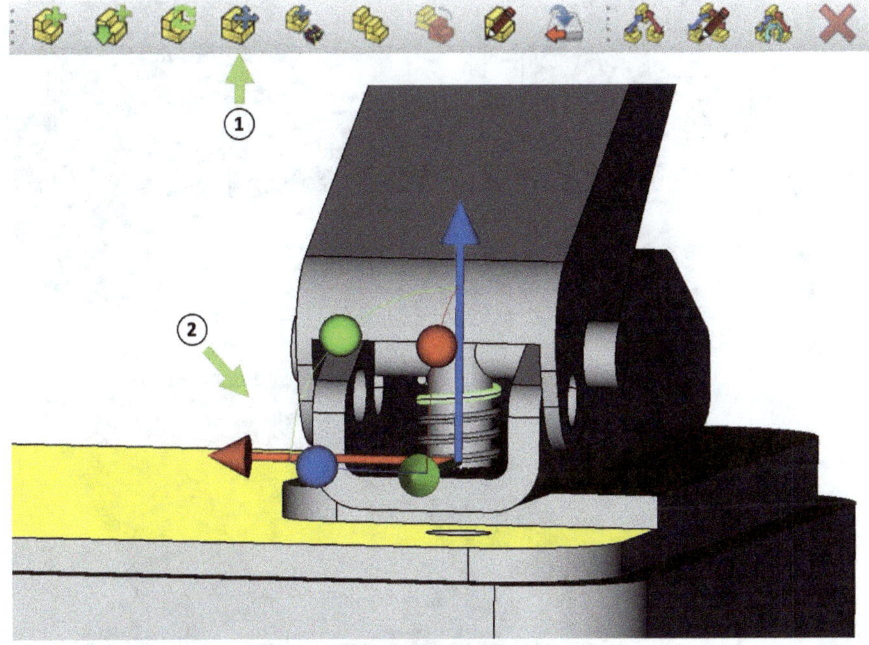

Then we can insert a second bolt into the assembly and link it just like the first one so that one of the two sides of the hole punch is fully assembled.

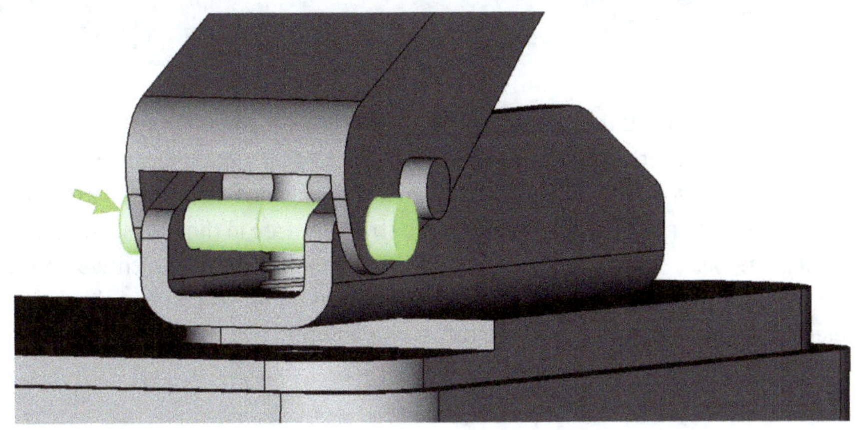

The components for the other side of the hole punch are linked in the same way. You can certainly do this on your own. The finished assembly should then look like this.

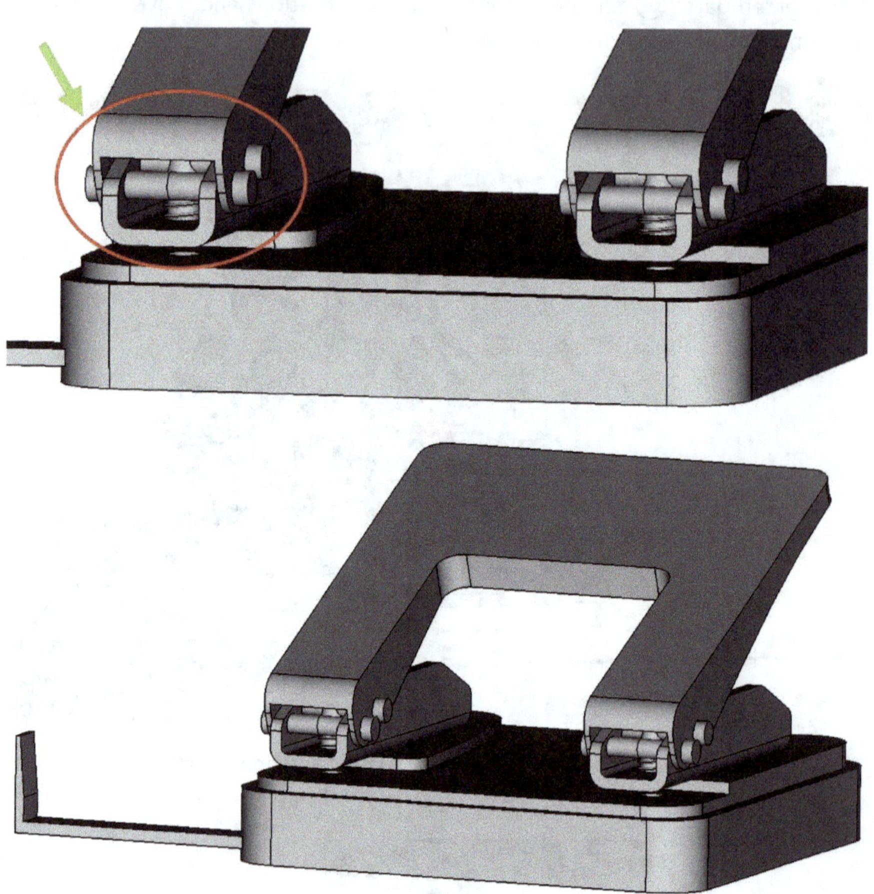

You may have noticed that we have not made any color changes to the individual parts. The hole punch looks a little dull in gray, but that's not a problem because we can also assign a different color to each of the individual parts in the assembly. We do this by selecting a component in the structure tree in the combination view — e.g., the base — and right-clicking on it. A menu opens, in which we choose "Appearance...". In the settings, search for the term "Shape color" and click on the selection button next to it. Here we can assign any color for the part.

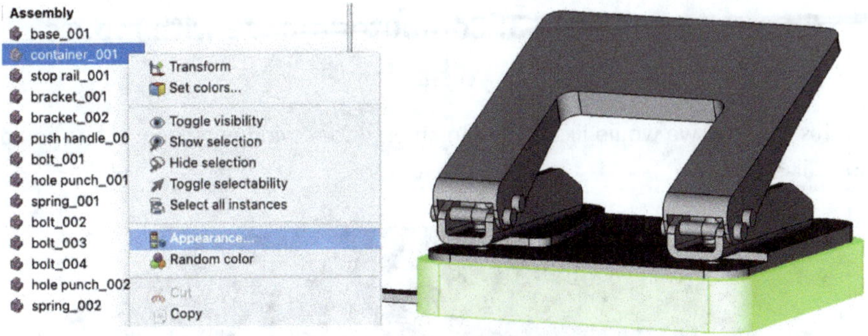

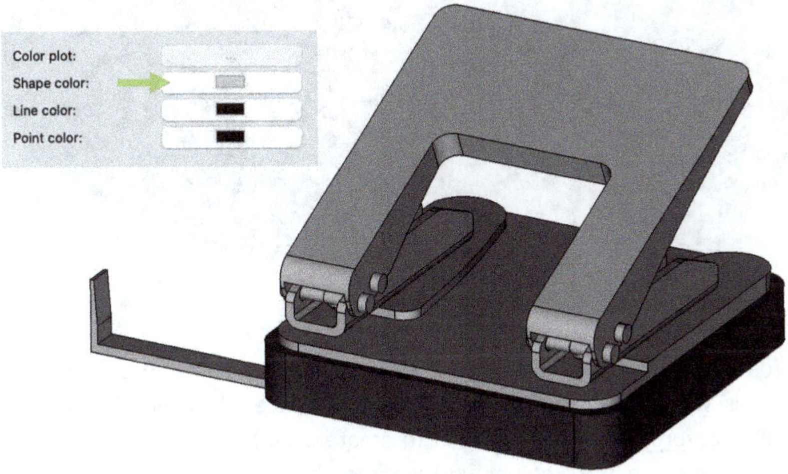

You can now assign a different color to each part according to your preference, adding a personalized touch to the hole punch. Perfect!

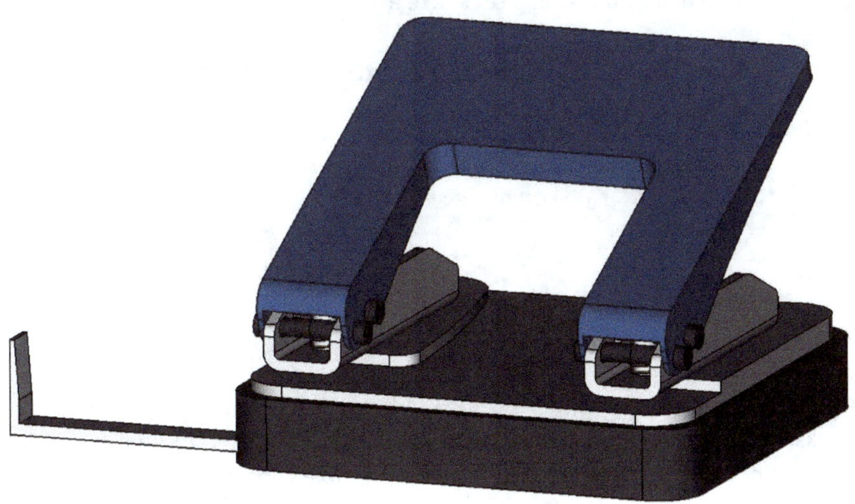

Chapter 4 | Project 3: Computer mouse with mouse wheel

In this chapter, we would like to design an ergonomic computer mouse. It should look like this.

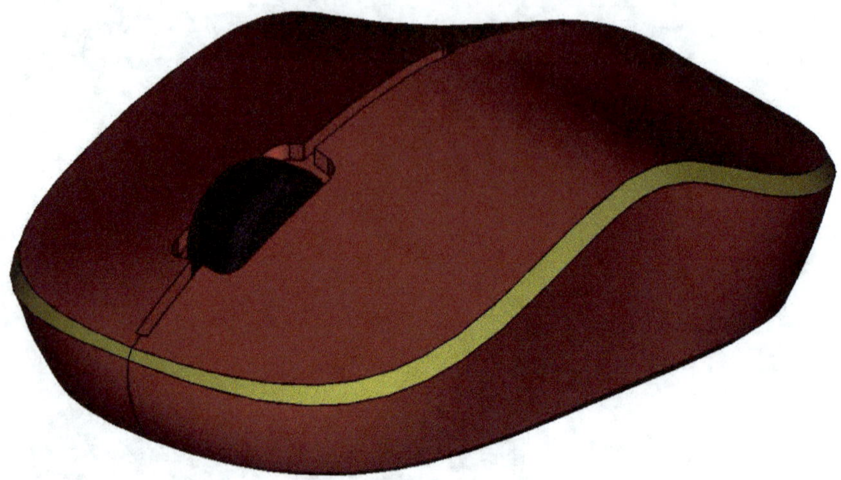

To create such an ergonomic mouse, we need to work with curves and surfaces. To do this, we start — <u>unlike usual</u> — in the workspace "Sketcher" ①. In this case, — <u>unlike before</u> — we do <u>not</u> need to create a body first. Similar to the "Part Design" workspace, the sketcher workspace itself also offers the option of creating sketches ②. Procedure: We first create several sketches to create a wireframe (outer contours) of one half of the computer mouse. We then add the surfaces and get the second half by mirroring. Let's take a look!

4.1 The basic body of the PC mouse

We create the first sketch on the x-y plane.

After selecting the command, "FreeCAD" automatically remains in the "Sketcher" workspace, only the menu bar changes so that we can find the usual commands for creating sketches. In this sketch, we create a curve geometry with the command

"B-spline by control points" ①. After selecting the command, we set the control points ② - ⑥ — starting at the origin ② — freehand approximately as shown and press the ESC key after the last point to end the command.

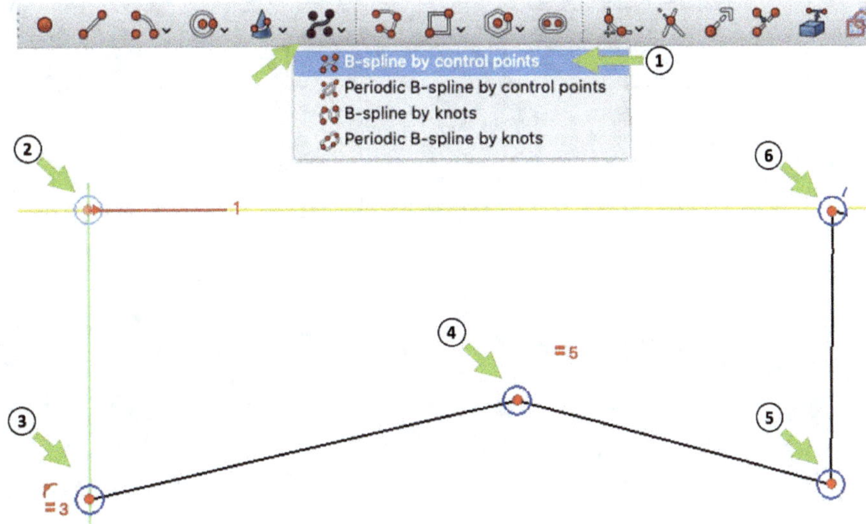

Executing the command generates the following curve based on the set control points. It is sufficient if the curve only looks roughly similar at first.

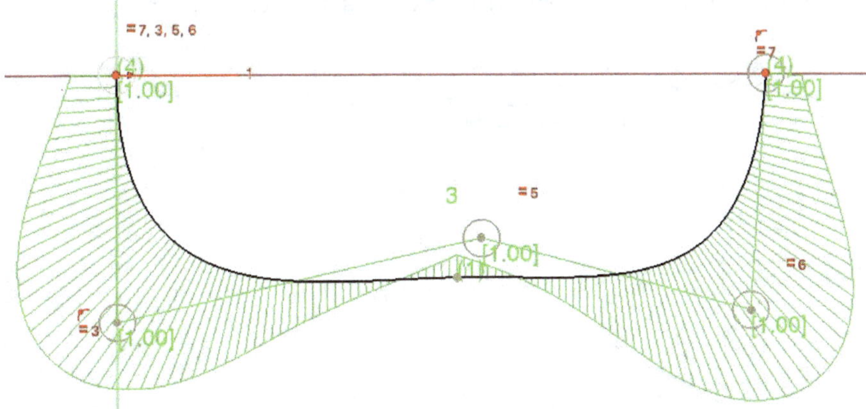

We now obtain the complete definition of the sketch by dimensioning the control points. Before dimensioning, we set a vertical constraint ③ between the two points ① and ②. Please dimension the control points as shown. You can also move the control points a bit if the sketch changes adversely during dimensioning.

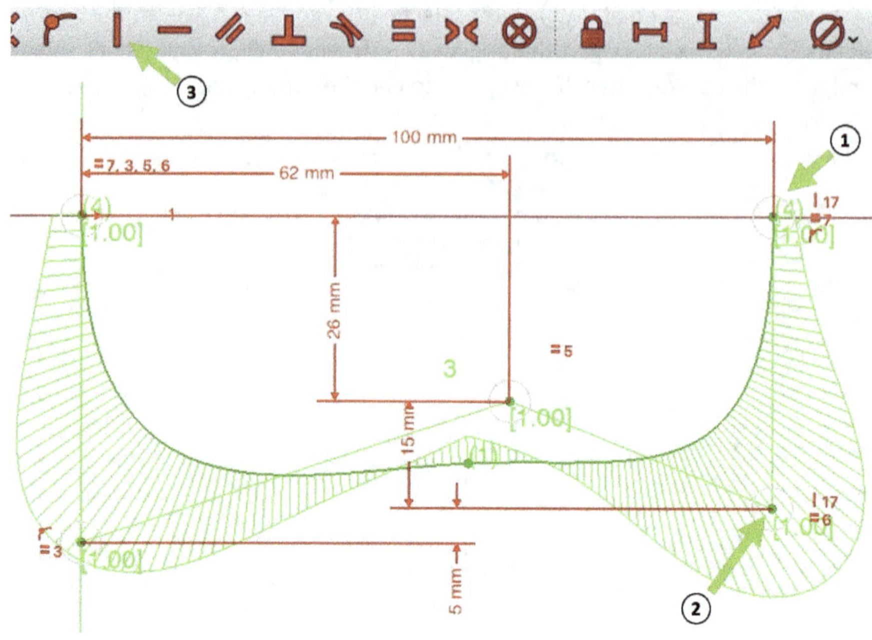

We can then close the sketch with the ESC key and create another sketch, but this time on the x-z plane. In this sketch, we us the same command ① as before (press the ESC key after the last point). Before we dimension, we set a horizontal constraint ⑥ between the points ② and ③ and between the points ④ and ⑤.

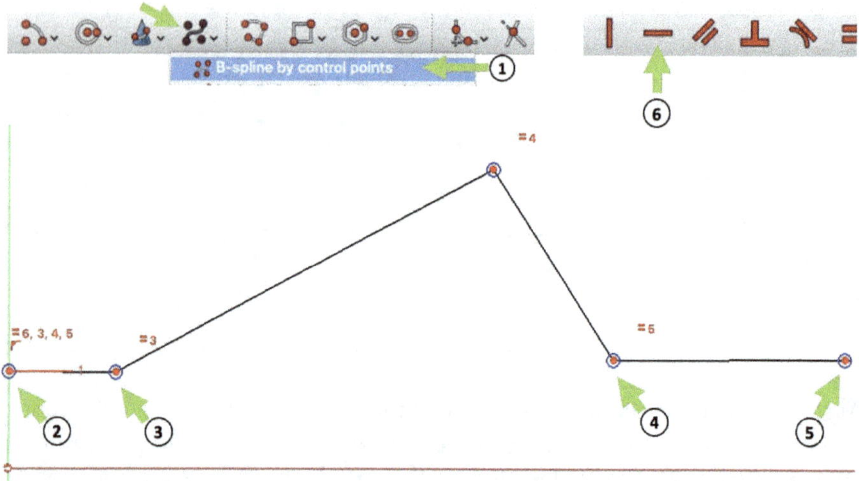

The dimensioning of the control points should be as follows.

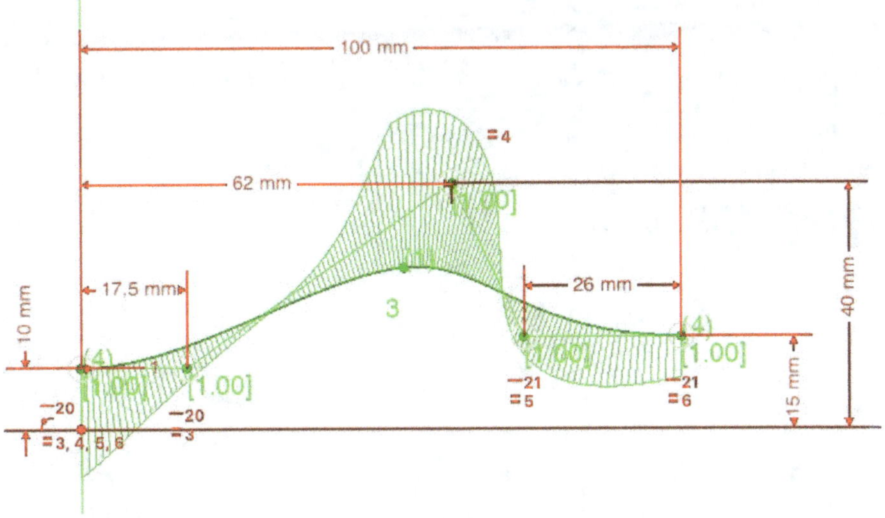

When the sketch is fully defined, we can close it with the ESC key. Let's save the document created so far. Before we can continue, we must first install another workspace. We do this via the "Addon manager", which can be found in the menu bar under "Tools". In the "Addon manager" search for "Curves workbench", click on it and then select the button "Install".

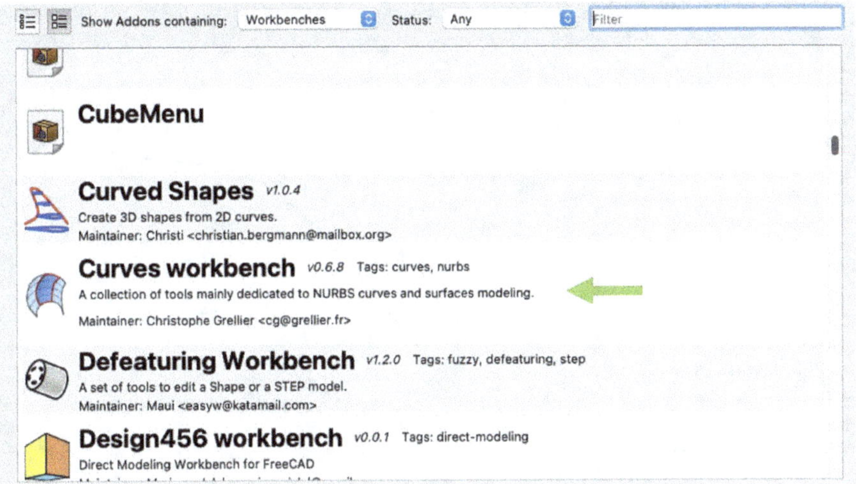

Close the "Addon manager" and restart "FreeCAD" for the changes to take effect. After the restart, we open our previously created document and switch to the "Curves" workspace ① that we have just installed. Here we first select both sketches ② and then create a 3D curve using the command "Mixed curve" ③.

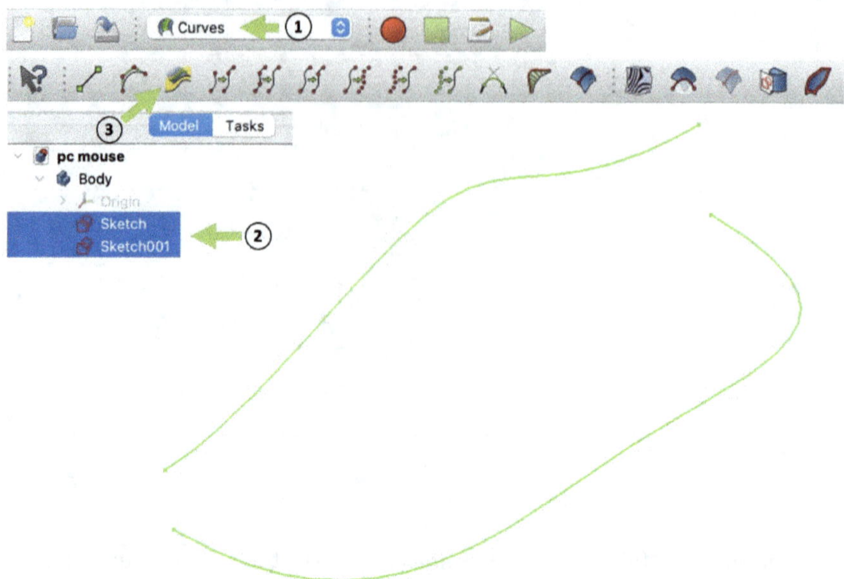

After selecting the command, the following 3D curve should be created from the intersection of the two previous 2D curves.

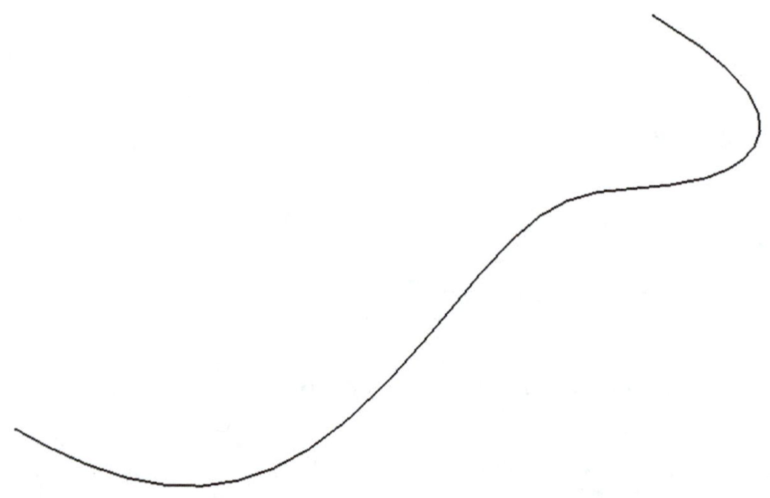

We then switch back to the "Sketcher" workspace ① and create a new sketch ② on the x-z plane.

In this sketch, we first use the familiar command "Create external geometry" ①
and click on the curve ② so that the geometry is projected onto the sketch.

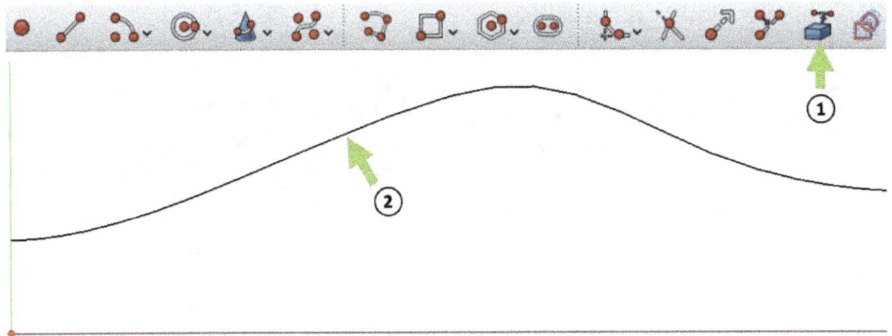

Afterwards, we create two 3-point arcs. We start the first arc at point ② and set
the end point to the x-axis. The second arc starts at point ③ and also ends on the
x-axis. Let's dimension the horizontal and vertical distance with 3 mm and 5 mm
respectively, and the radius of the two arcs with R13 mm and R25 mm.

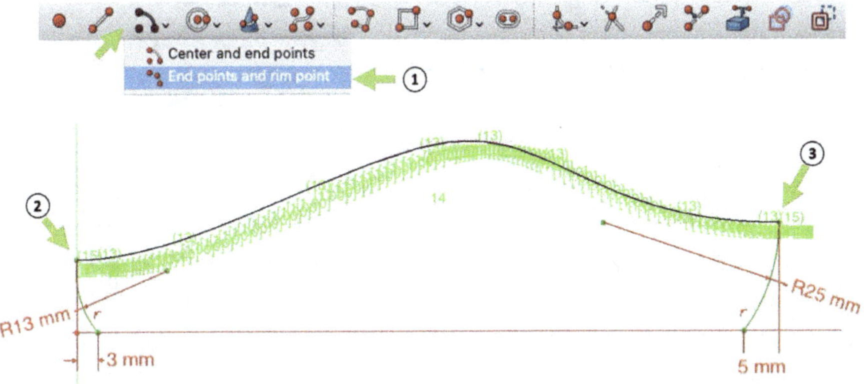

We then close the sketch with the ESC key and create another sketch — this time
on the x-y plane. In this sketch, we first project the horizontally displayed curves
② and ③ using the command ①.

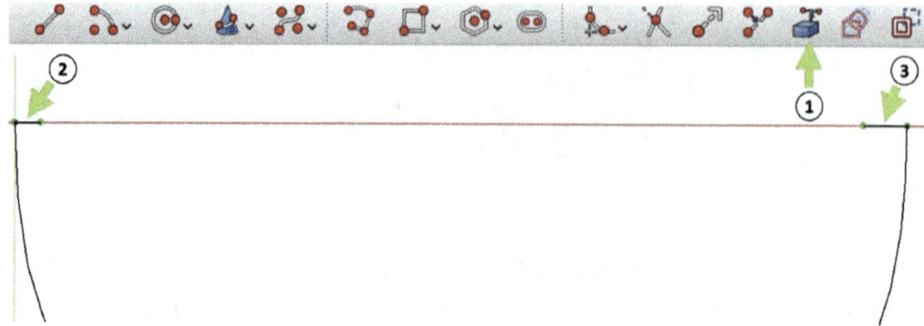

Please create another curve as shown.

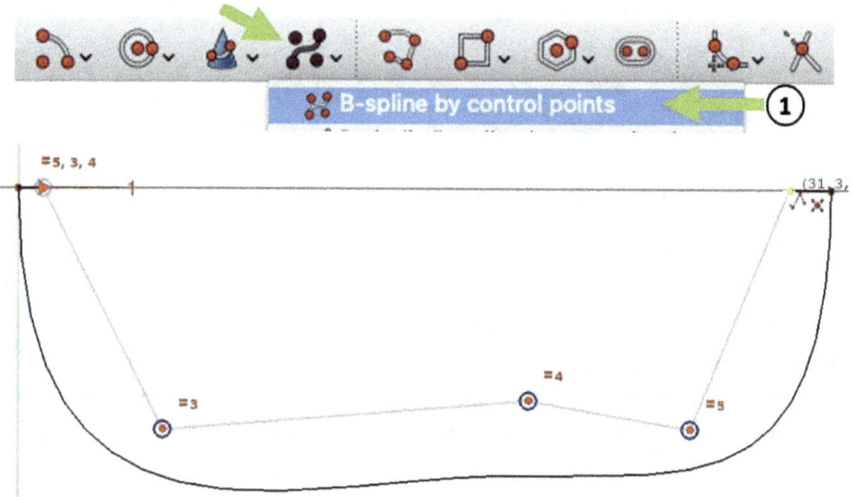

And dimension the control points as follows.

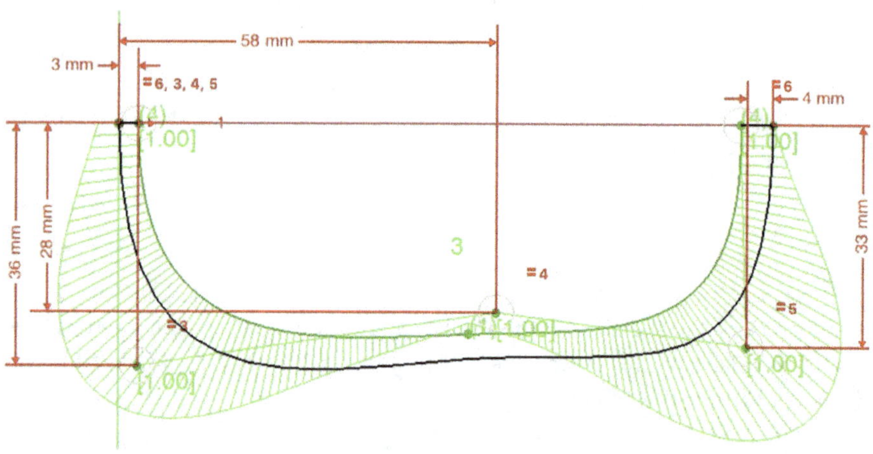

Now we can end the sketch with the ESC key. Subsequently, we start another new sketch on the x-z plane. The contour of the mouse (side view) can already be seen in the sketch. Start by projecting ① the curve ② into the sketch.

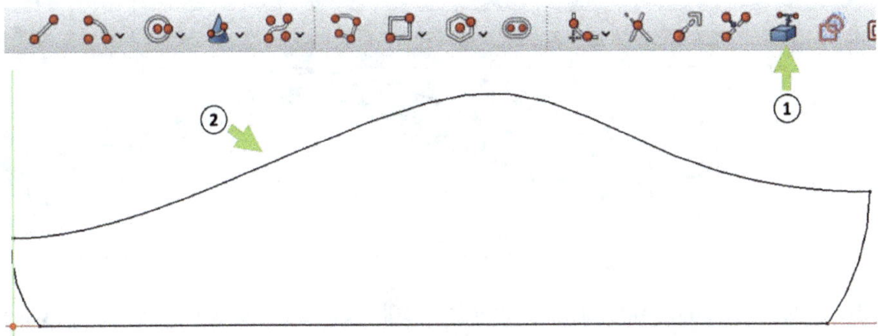

Now we can sketch another curve using "B-spline by control points" ①.

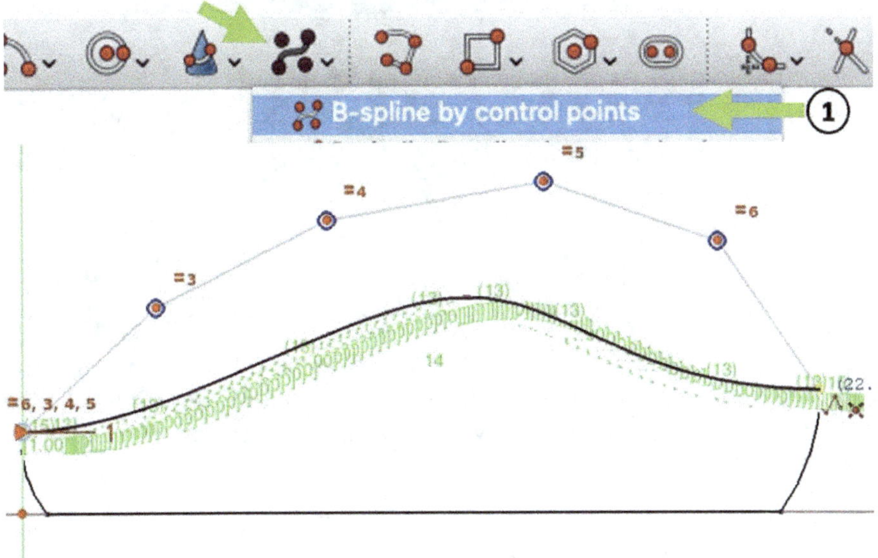

This curve will later represent the curvature of the surface of the mouse — here in the side view. You can either dimension this curve as shown below, or simply position the control points approximately.

Note: For better presenting the dimensions, the points are connected with lines in the following illustration. The curve has been removed here.

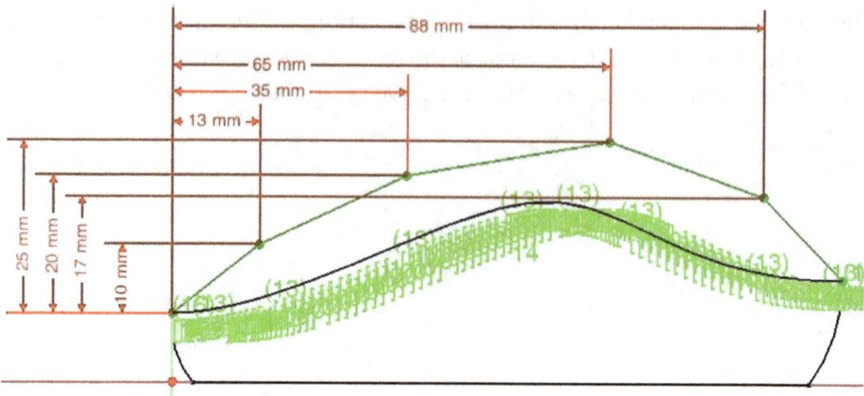

The following illustration shows how it should look in your window (curve with control points; dimensions are identical to the previous illustration).

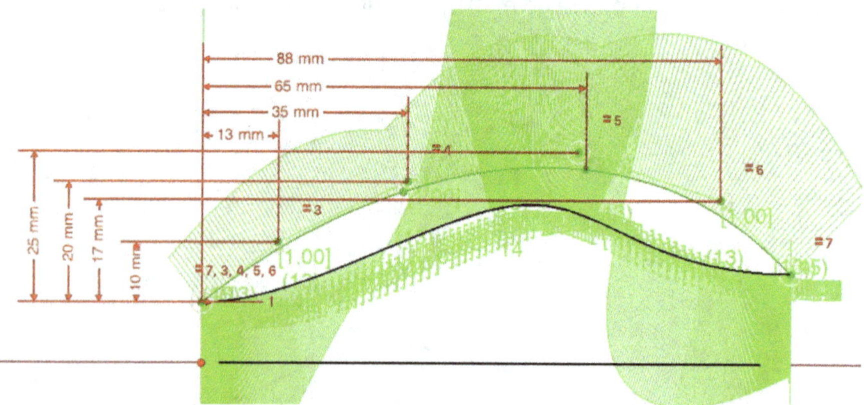

After dimensioning, you can close this sketch with the ESC key. Great! All the relevant sketches, forming a kind of wireframe model, are complete. The frame of one half of the computer mouse is done.

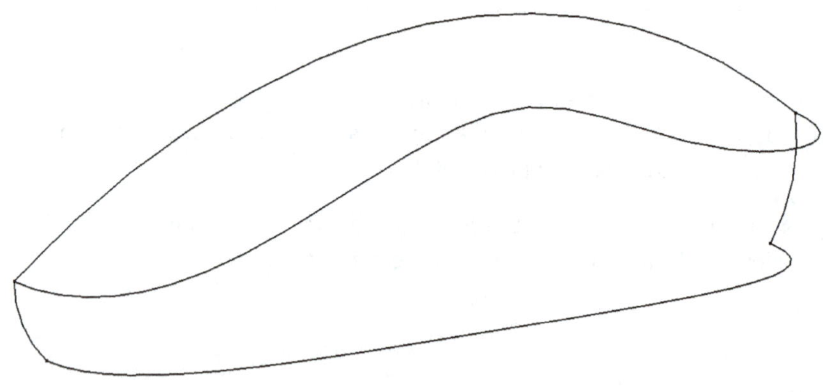

Next, we can create the surfaces. To do this, we switch to the workspace "Curves" ① and select the curves ② - ⑤ one after the other (CTRL key pressed). Lastly, click on the command "Gordon surface" ⑥.

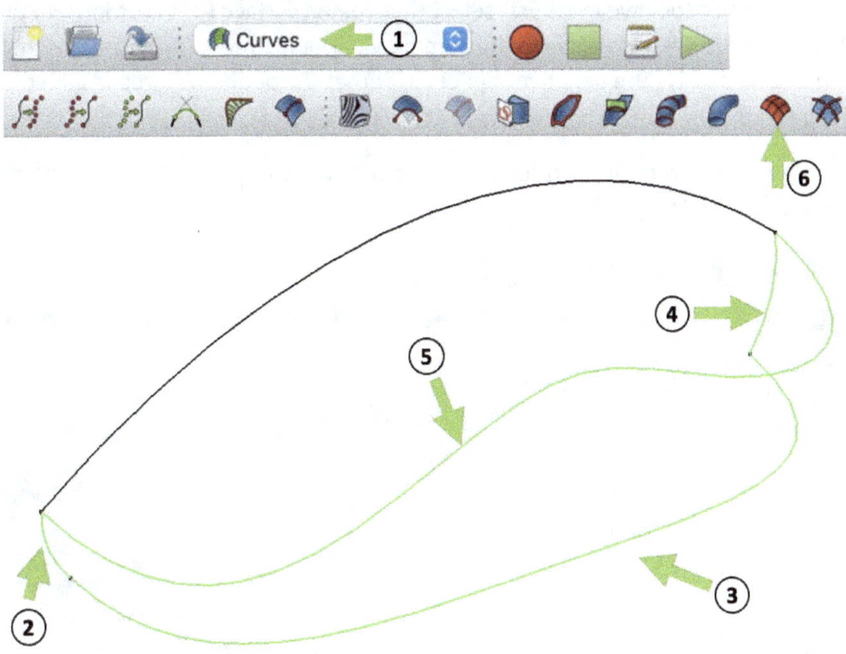

In this way, the program creates a surface within the curves.

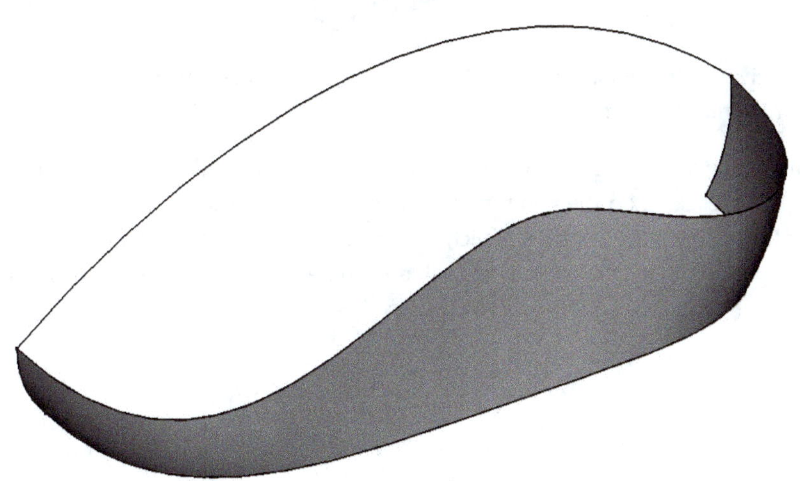

But such a thin surface is not enough for us, we want our mouse to have a wall that has a certain thickness.

Therefore, we switch to the workspace "Part" ①, in which we use the tool "Offset" ③. To use this tool, we must first select the created surface ② ("Gordon") in the structure tree. Then we click on the command ③ and enter a distance of -2 mm in the settings ④. The negative sign ensures that the additional area is added to the outside.

To create a wall from the two surfaces, we need to activate the option "Fill offset" ⑤ in the next step.

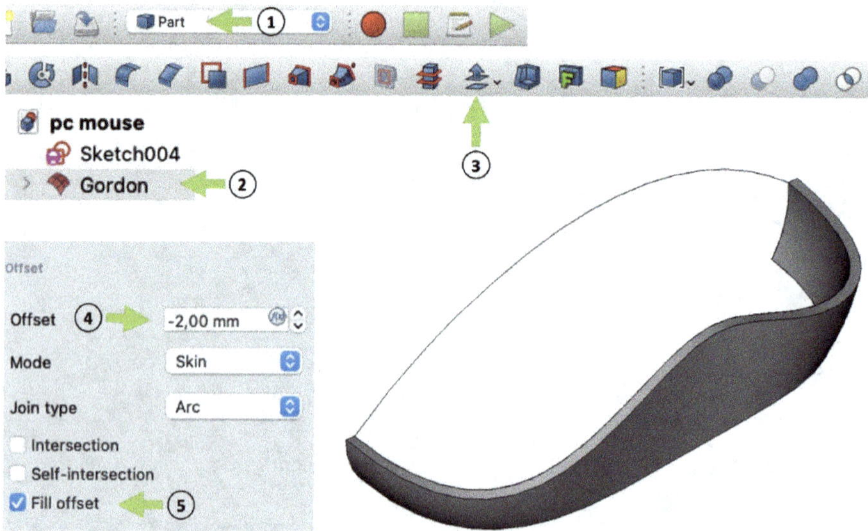

That gives us the first side panel of the mouse. We can create the second side panel by mirroring the first one.

Simply do this by selecting the first side panel ① ("Offset") in the structure tree and then clicking on the command "Mirroring..." ②. In the last step, we need to set the correct mirror plane, which in our case is the x-z plane ③. Confirm with "OK" and the second side wall will be created automatically. This is a great command!

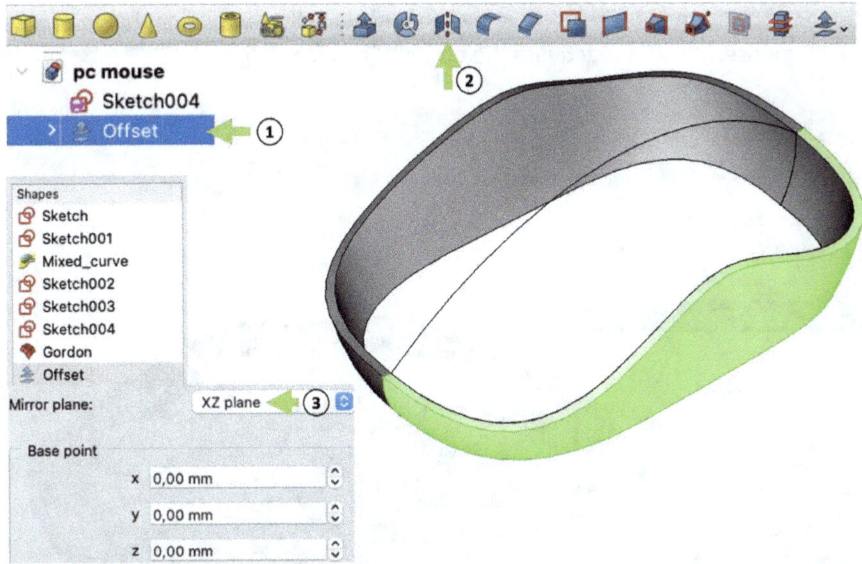

We then switch to the "Surface" ① workspace, in order to create the upper cover surfaces of the mouse. Here we use the command "Sections" ②, which can create a surface from a series of edges. Once we have clicked on the command, we need to click on "Add Edge" ③ in the settings and choose the two edges ④ and ⑤ one after the other.

We also want to create a wall from this surface. To do this, we need to switch back to the workspace "Part" ①, select the surface we have just created ("Surface") ② in the structure tree and use the command "Offset" ③ as before. We select a distance of +2 mm ④. The positive sign should place the surface on the outside of the PC mouse. We also reactivate the option "Fill offset" ⑤.

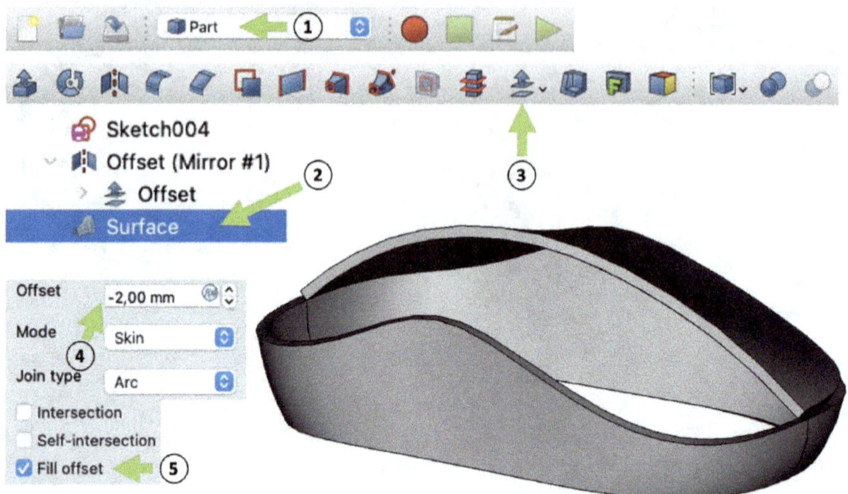

Similar to the side wall, let's use the command "Mirroring..." ② to add the second top wall. First select the wall you have just created ①, then click on the command ② and set the mirror plane to the x-z plane ③. After confirming with "OK", the PC mouse should look as shown.

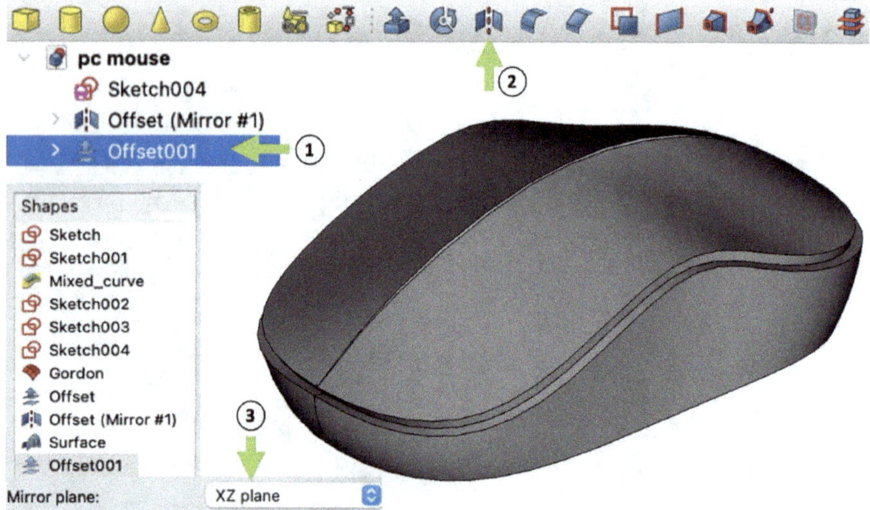

In the next step, we want to fill the gaps in the upper area and the upper edges. We do this with the command "Ruled Surface" ③. To apply the command, first click on the two edges ① and ② (CTRL key pressed) and then on the command ③. This will fill the gap with a surface.

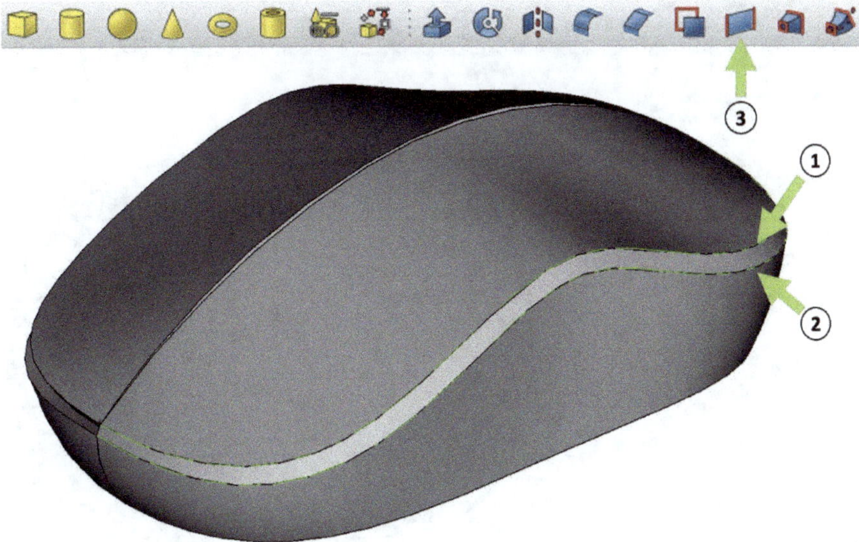

We also apply this command to the opposite side and the upper area one after the other.

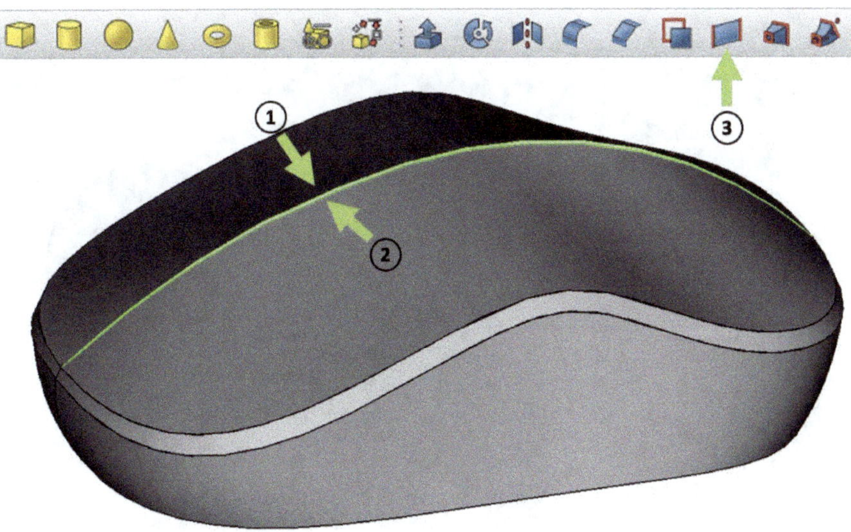

The only thing missing for the base body is the bottom surface of the mouse. We create it in a similar way to the top surfaces. To accomplish this, we first switch to the workspace "Surface" ① and use the command "Sections" ②. We select the two inner edges ④ and ⑤ on the underside of the part as edges.

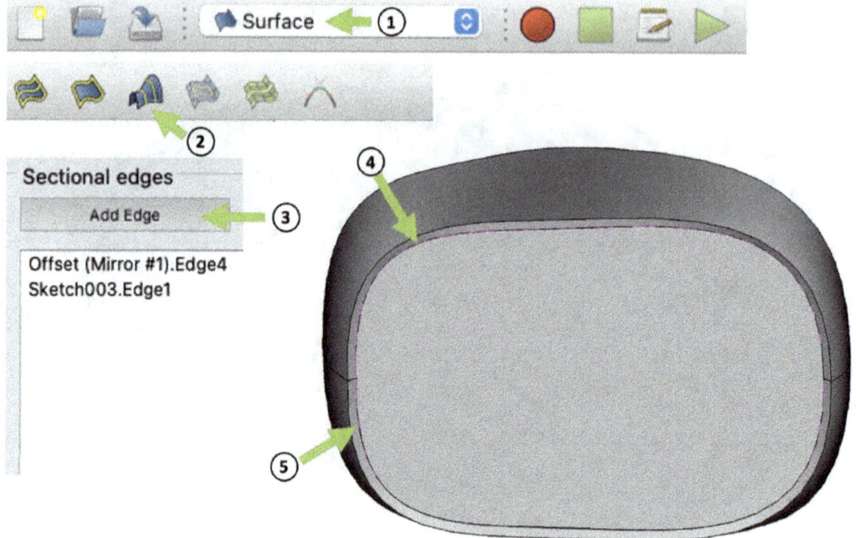

We then switch back to the "Part" workspace and apply the "Offset" command once again with a value of 2 mm (① - ⑤).

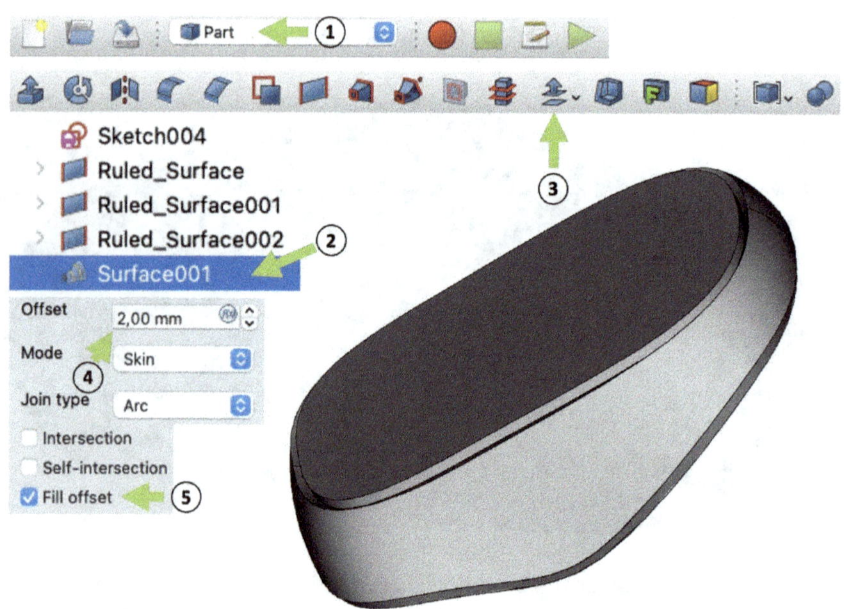

The resulting gaps are subsequently filled by using the command "Ruled Surface". The procedure is identical to that described above in the upper part of the section.

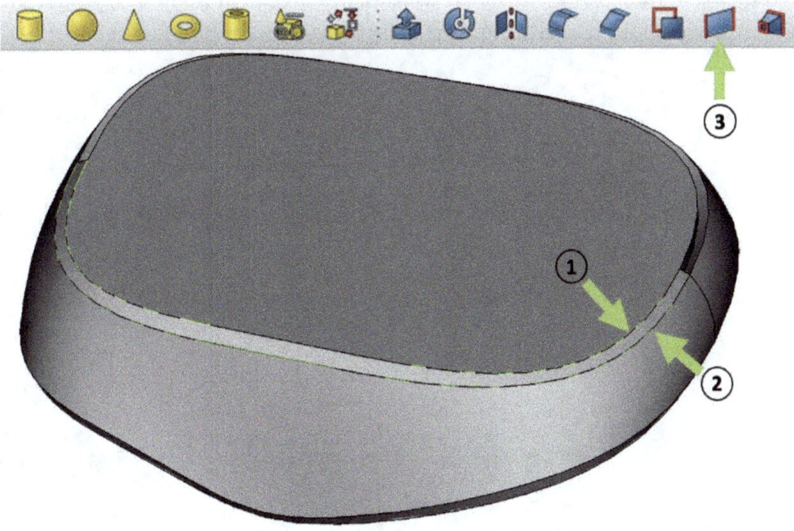

After we have applied this command to the opposite side as well, the basic body of the PC mouse is finished. Perfect! Now we want to create the buttons and the mouse wheel. To achieve this, we create a cut-out on the top of the base body. We create a sketch (① - ③) and switch to the section view ④ using the command "View Section".

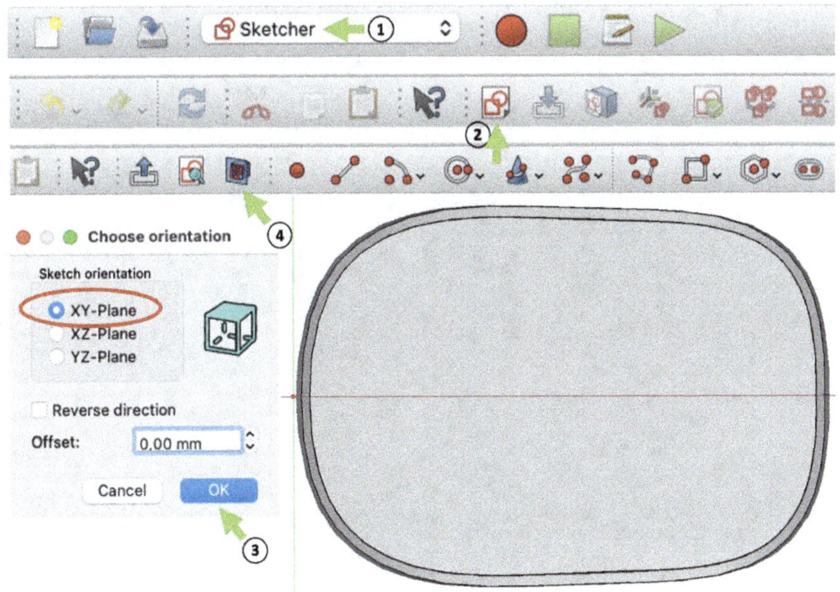

In this sketch, we draw the following profile with two rounded corners. We use the familiar commands "Create Polyline" ①, "Constraint-preserving sketch fillet" ③ and "Constrain radius" ⑤. It is best to start the profile at point ②, which should be 0.5 mm horizontally and 1 mm vertically from the coordinate origin.

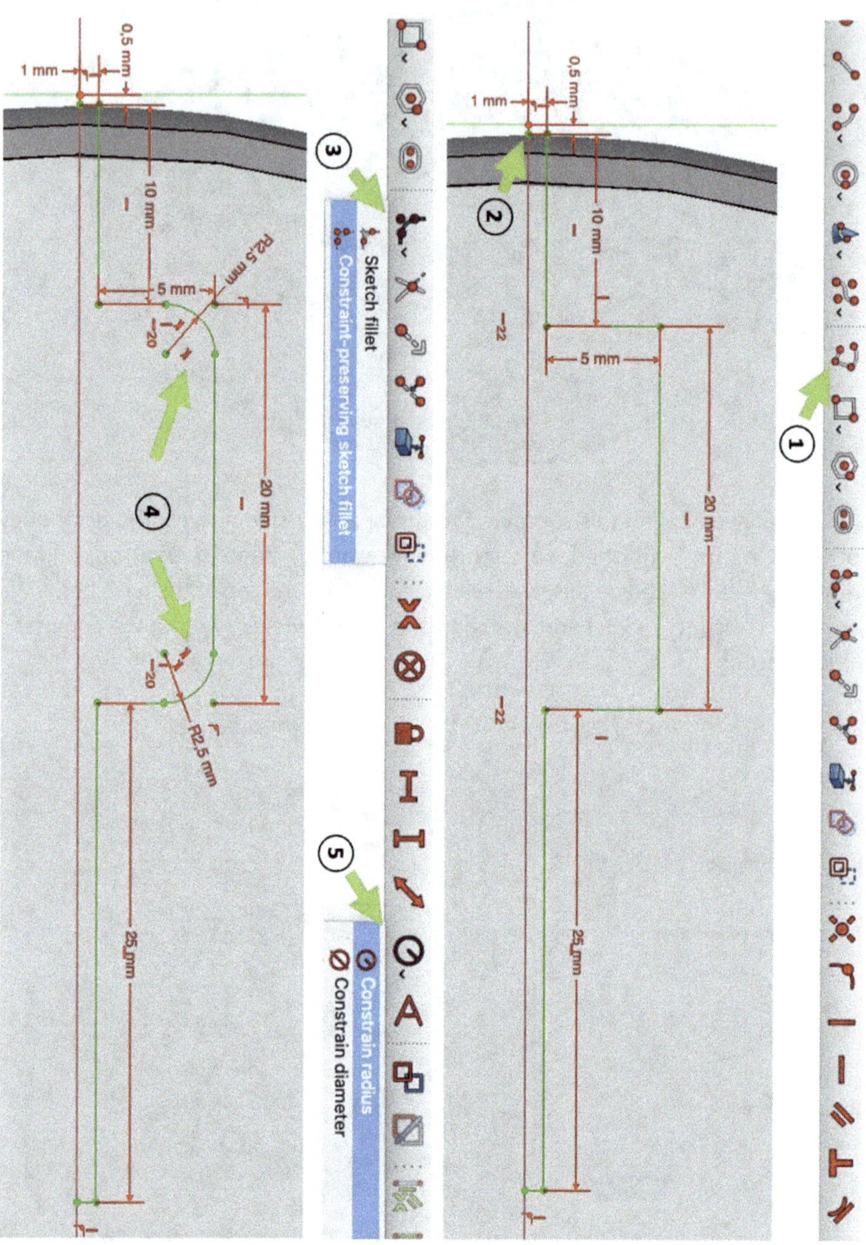

The profile just sketched represents one half of the section. For the second half, we mirror the geometry on the red x-axis. We can do this by first clicking on all the elements that we want to mirror ① one after the other. Then we also click on the mirror axis (x-axis) ② — the axis must always be selected last with this command — and then click on the command "Symmetry" ③. This gives us the complete profile.

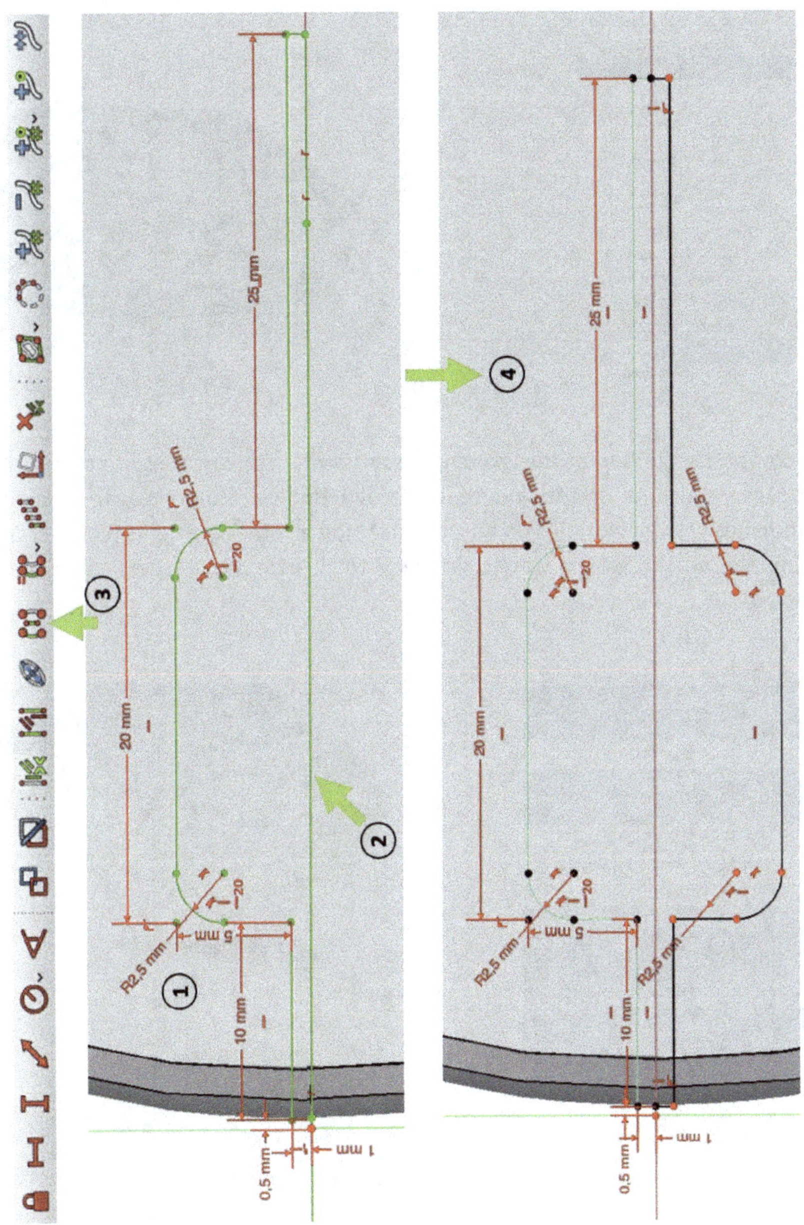

Please close the sketch and click on it in the structure tree ①. If we click on the small arrows next to the parameters "Placement" ② and "Position" ③ in the "Base" area, we can move the sketch. In this case, we need an offset of 40 mm in the z direction ④ so that the sketch floats above the mouse. We could also have made this setting when creating the sketch, e.g., by creating a parallel plane for the sketch.

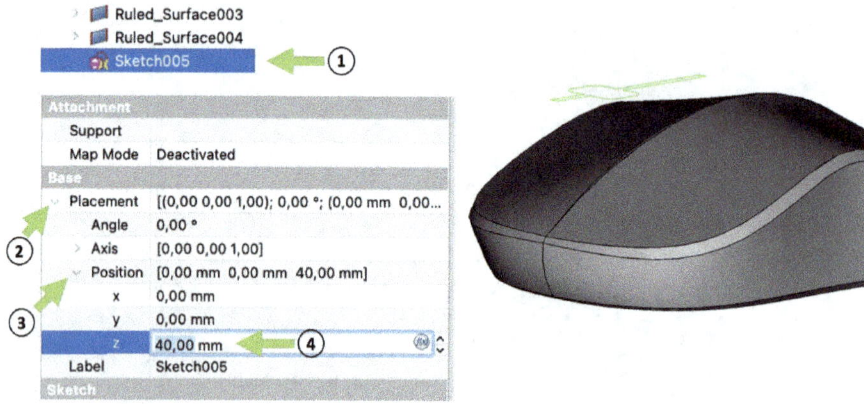

To create a section with this sketch, we must first convert the two upper cover surfaces / buttons of the mouse into solid bodies. Otherwise, the following commands will not work. To do this, we first hide all other elements (colored green or blue here). The easiest way to do this is to click on the 3D parts and press the space bar.

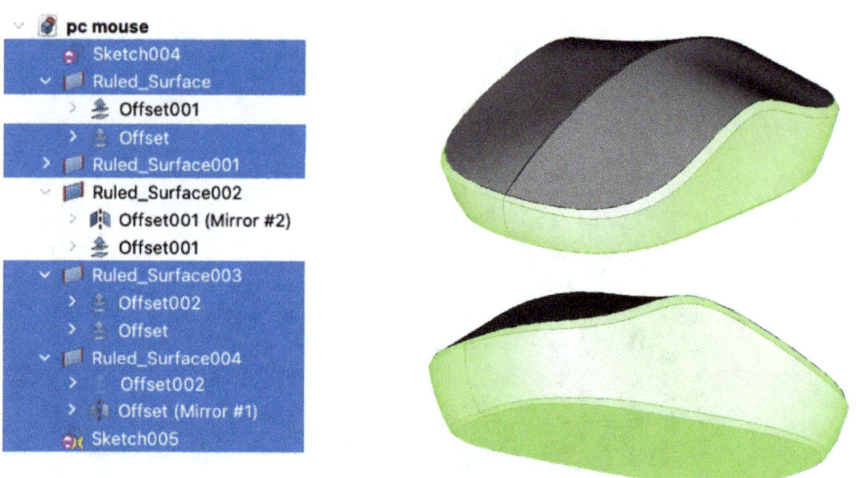

Only the following three elements should now be displayed ①.

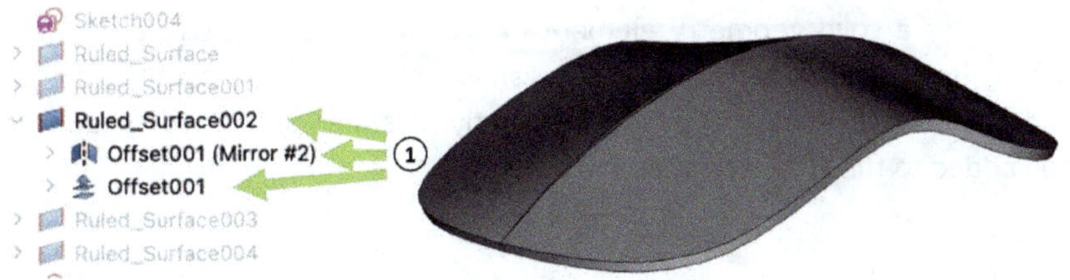

To turn the buttons into solid geometry elements, we switch to the workspace "Part" ① . Here we are going to use the command "Shape Builder..." ③ . First, however, we even hide the two other elements ② (space bar) so that only one button remains visible.

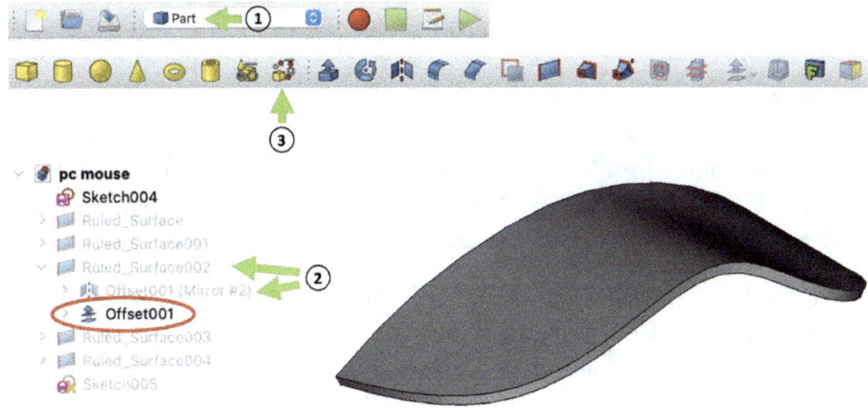

Now click on the command ① and select "Solid from shell" in the settings. We then select <u>all</u> faces of the button (CTRL key pressed) ③-④ and click on the command "Create" ⑤ afterward.

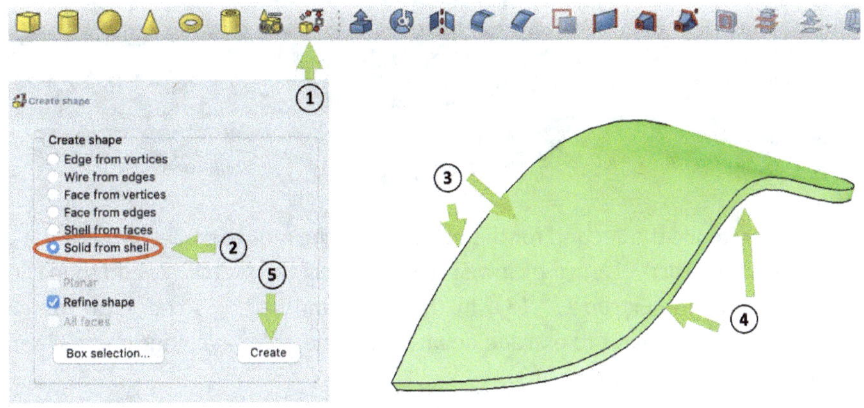

This creates a solid geometry element in a <u>background process</u> (it seems as if nothing is happening). However, we can simply close the command "Shape Builder...". If the command worked correctly, the element "Solid" should have been added to the structure tree.

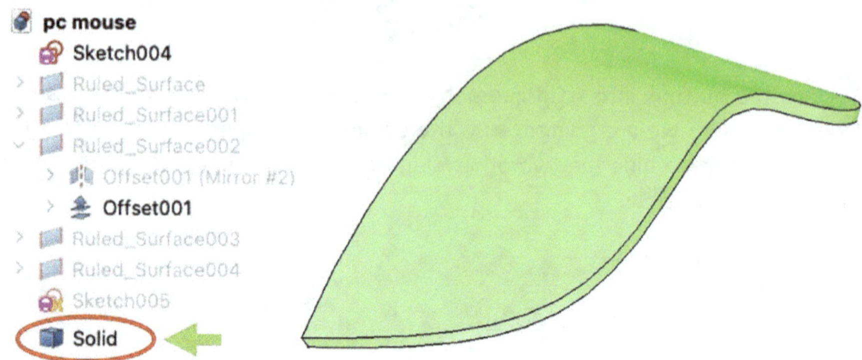

We use the same procedure for the other button. To do this, we first hide the "Solid" and "Offset001" we have just created with the space bar and then show "Offset001 (Mirror #2)" with the space bar.

If the command also worked for the other button, "Solid" and "Solid001" ① should now be listed in the structure tree. We can then show all other elements — apart from ② - ③ — again (also note the sub-elements).

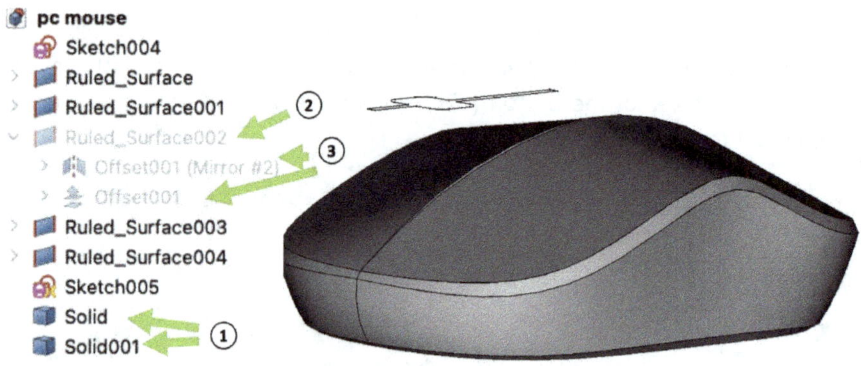

Now we can make the section for the buttons and the mouse wheel. We are still in the workspace "Part" ①, in which we now first select the sketch ② and then click on the command "Extrude..." ③. In the command settings, check the option "Reversed" ④ (we want to extrude downwards) and enter a dimension of 35 mm in the option "Length - Along" ⑤.

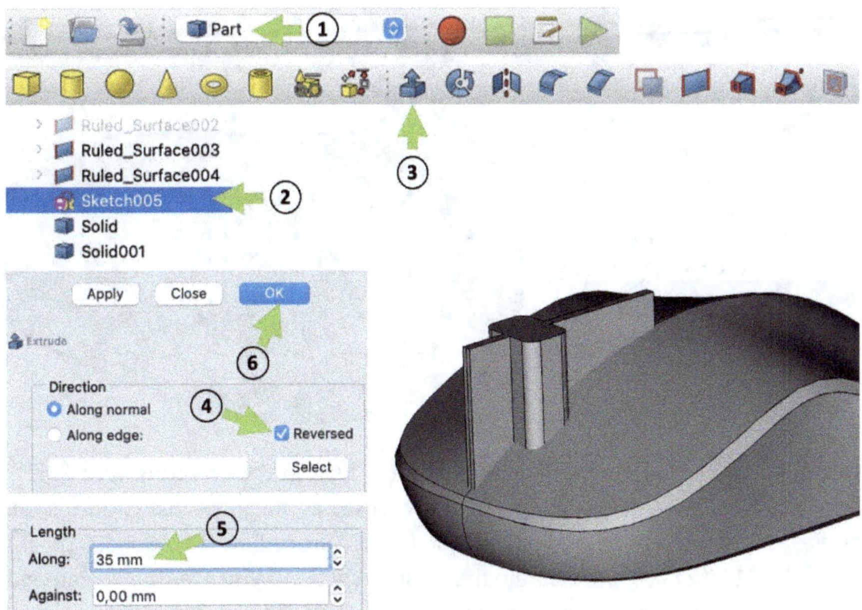

Now we need to subtract the extruded construct once from each of our two mouse buttons, and therefore need it twice. We copy the extrusion ① by selecting it and pressing CTRL+C. A window appears, which we confirm with "OK" ②. To paste the copied element, press the CTRL+V key command. This is a simple way of doubling the extrusion "Extrude" and getting "Extrude001" ③.

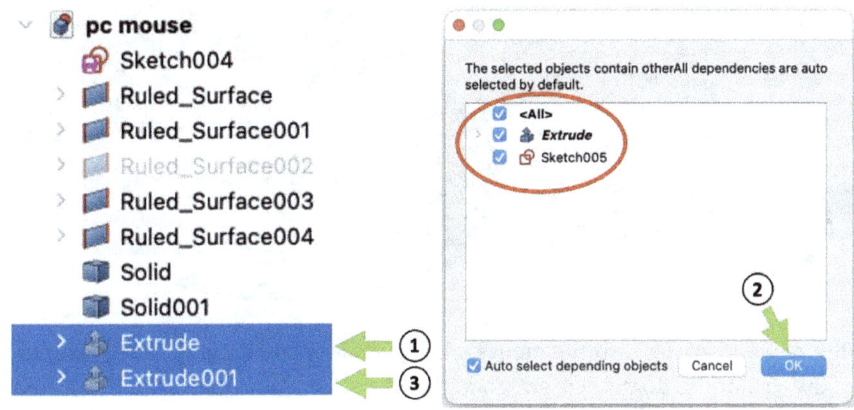

Now we can finally make the cut-out. We do this by first clicking on the solid geometry element of the first mouse button ①, then on the first extrusion ② and finally on the command "Cut" ③.

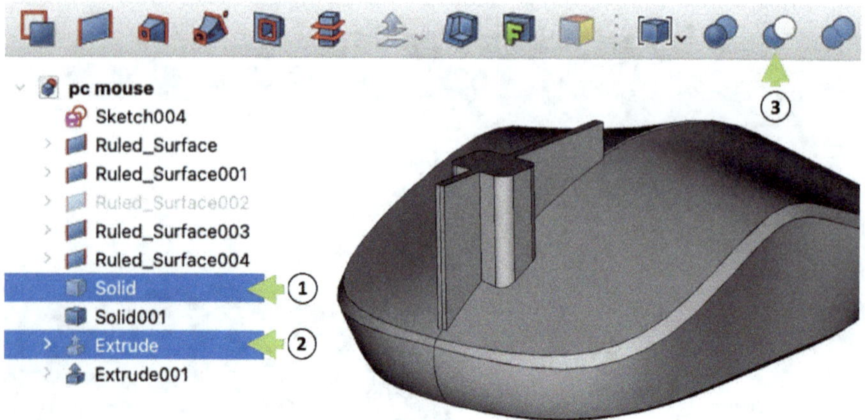

This merges the two geometry elements into "Cut" in the structure tree ①. At first, we do not see any change to the mouse, as the other extrusion is still displayed. After we have applied the same procedure to "Solid001" and "Extrude001" (② - ④), we obtain the finished section. With the "Cut" command, the order of selection is significant. In this case, the solid geometry element must be selected first and only then the extrusion. Try it the other way round and you will see why. However, it is better to save the file first.

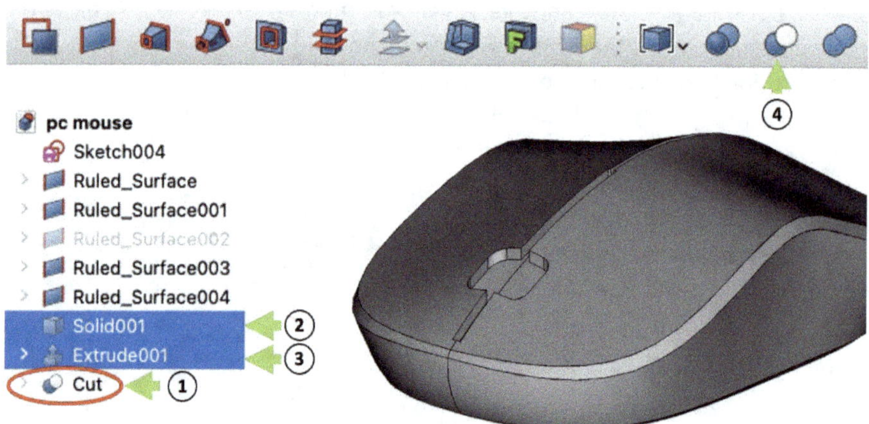

Before we continue with the mouse wheel, we need to fill a gap in the upper area of the section. To do this, we switch to the workspace "Surface" ① and first create a boundary curve ⑤. We can get this curve by selecting the two edges ② and ③ (holding down the CTRL key) and then clicking on the command "Blend Curve" ④.

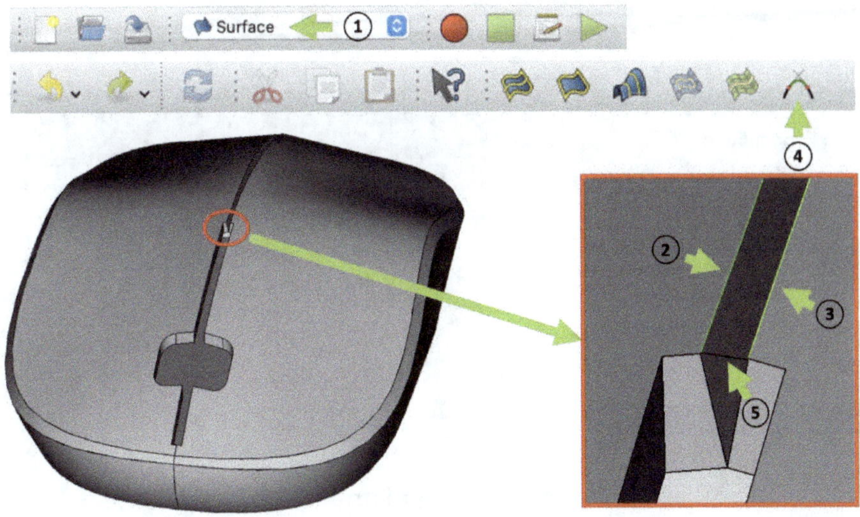

In the next step, we create a triangular filling surface ④ by clicking on the command "Filling" ① and then selecting the edges ② and ③. Confirm the command with "OK".

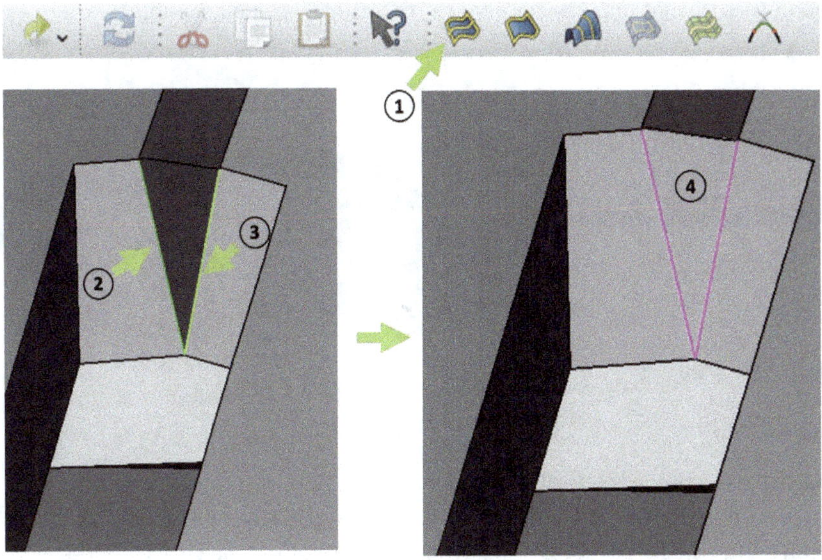

The last steps consist of switching to the "Curves" workspace ① and displaying the top surface of the gap ② (space bar). While holding down the CTRL key, we first click on the upper cover area ③ and then on the curve created earlier ④. Lastly, we click on the command "Trim face" ⑤, which cuts off the top face at the boundary curve so that the overhang is removed, and the gap is closed.

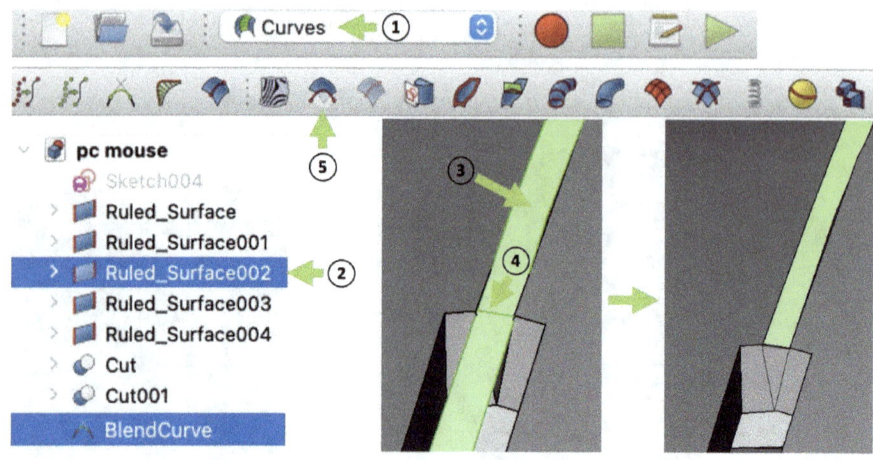

4.2 The mouse wheel and its holder

Now let's get to the holder for the mouse wheel. We want to create this on the inside of the base element of the mouse. To be able to do this, we first need to create a solid element from this surface base element. We do this in a similar way to the two mouse buttons in the previous chapter. First, we hide all other geometries so that only the base (here "Offset002") remains visible.

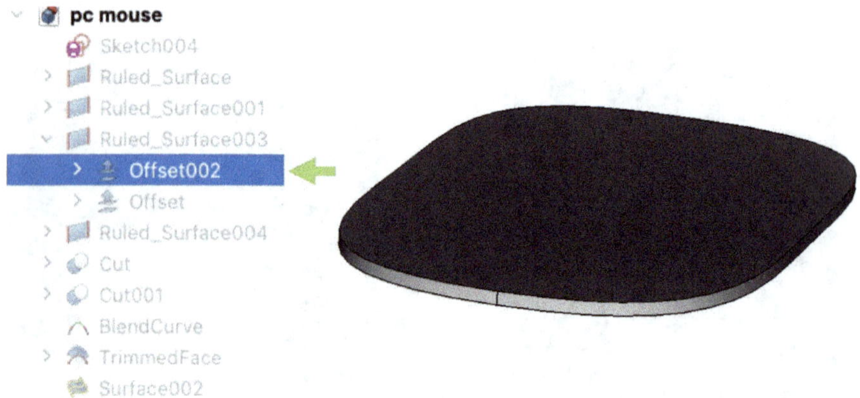

Please try the process on your own. If you need help, you can take another look at the procedure for the buttons. After the command has been executed, you should receive the object "Solid002" (or similarly named) ① in the structure tree.

We now need a sketch on the top of the base for the base body of the bracket. To do this, we switch to the workspace "Sketcher" ②, click on the top side ③ of the part and then select the command "Create Sketch" ④.

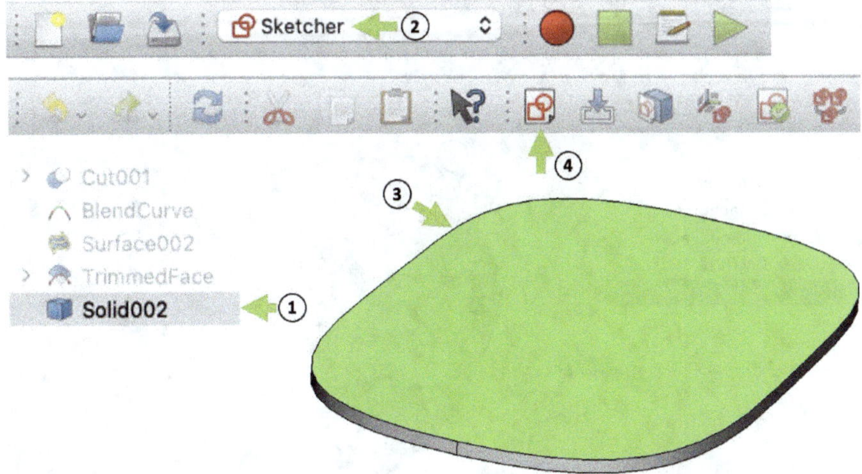

In this sketch, we draw two rectangular profiles with the following dimensions. The reference point for the position dimensions is — as often — the coordinate origin.

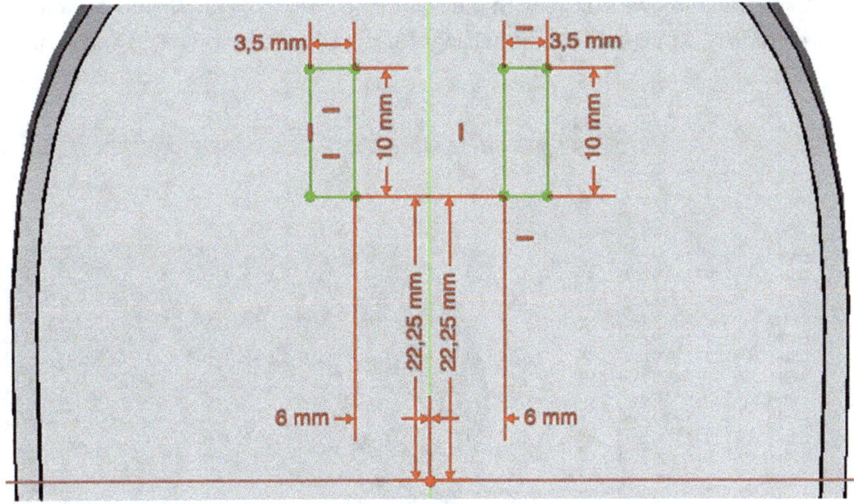

After we have closed the sketch, we switch to the workspace "Part" ① and use its function "Extrude..." ③. Before we click on the command, we select the sketch we have just created ②. In the settings, we enter a dimension of 20 mm ④ and can then click on "OK" ⑤.

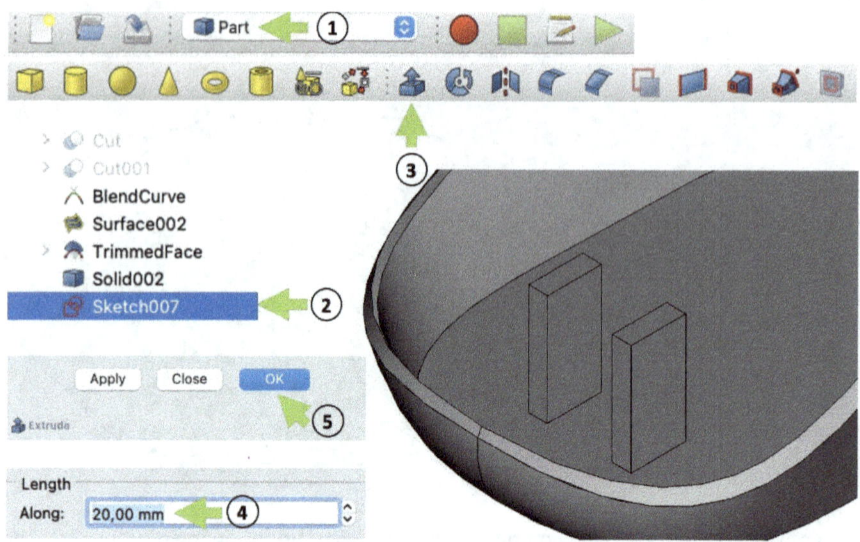

Now, we would like to round the upper edges ①. After we have selected these (CTRL key pressed), we do this with the command "Fillet..." ②. We must enter the desired radii (2.5 mm each) in the settings for the selected corners ③ and confirm with "OK".

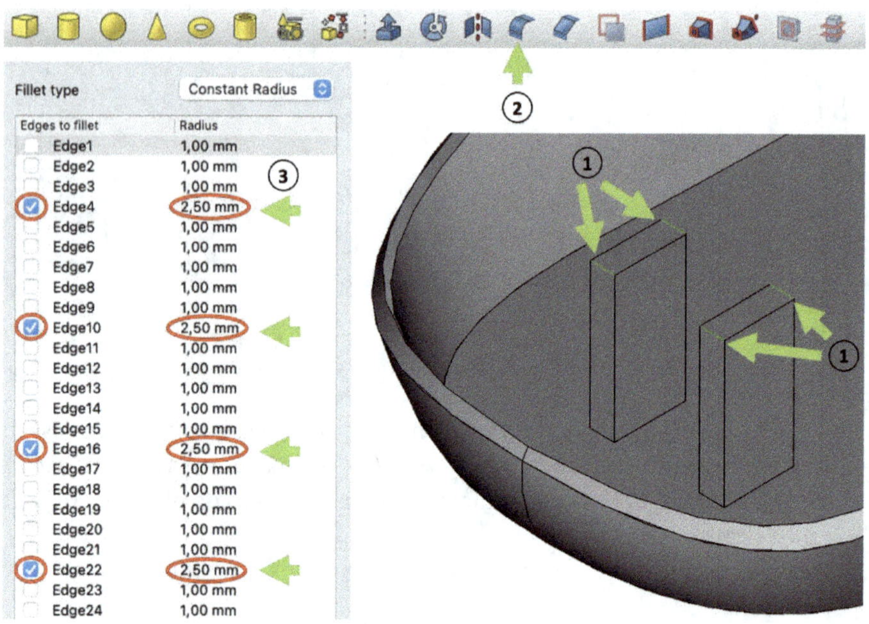

Excellent! Once we have saved the progress made so far, we can design the mouse wheel. We do this in a new document and — as usual with solid parts — in the workspace "Part Design". Here we can — as always for solid parts — create a body ② and a sketch ③ on the x-y plane.

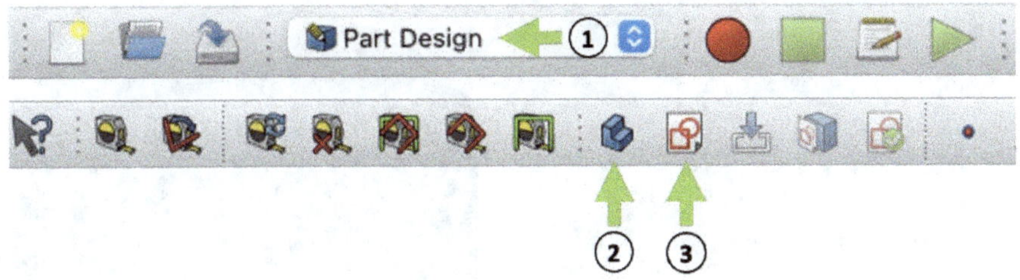

In this sketch, we create the basic body of the mouse wheel. First, we need a simple circle with its center at the coordinate origin and a diameter of 20 mm.

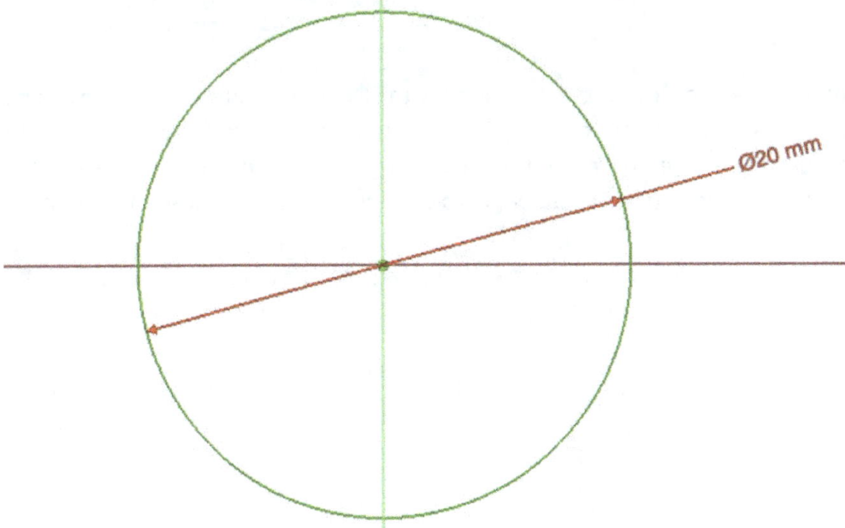

Next, we want to extrude the sketch. We do it — as usual in this workspace — by clicking on the command "Pad" ①. The program will most probably recognize the sketch automatically (if not, select the sketch).

In the command settings, we first switch to the option "Two dimensions" ② at "Type" to create a symmetrical extrusion in two directions. This places the x-y plane in the middle of the part, which is important for the later procedure. In the settings "Length" and "2nd Length", we want to enter 5 mm each (③ and ④) so that the part gets a total thickness of 10 mm.

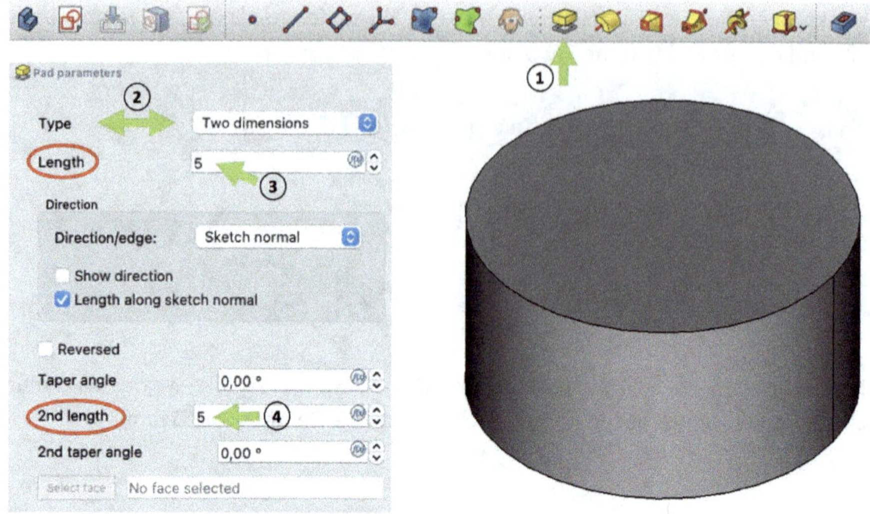

After we have executed the command with "OK", we create another new sketch on the x-y plane. Before drawing, it is best to hide the previous body (select it in the structure tree and press the space bar). We sketch a simple circle with a diameter of 2 mm (coordinate origin as center) and can then close the sketch.

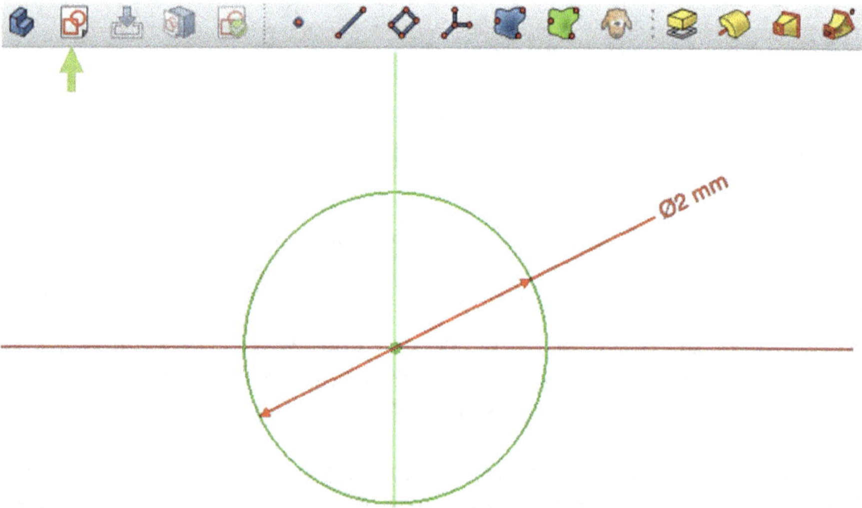

Now, we make the previous body visible again (space bar) and create another extrusion with the command "Pad" ①. As before, the sketch is selected automatically. In the settings, we choose "Two dimensions" ② for the option "Type" and a dimension of 7.5 mm in both directions (③ and ④).

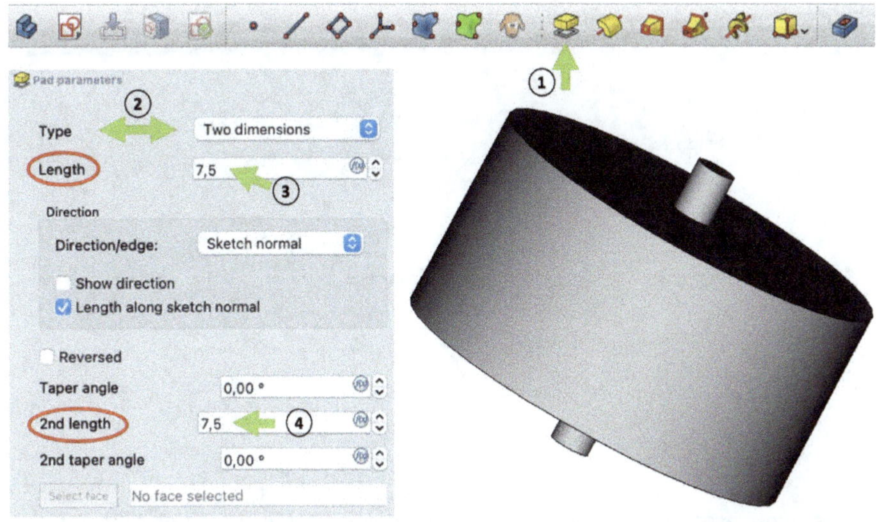

Additionally, let's round off the two edges ① and ② with the command "Fillet" ③ and a radius of 2 mm each.

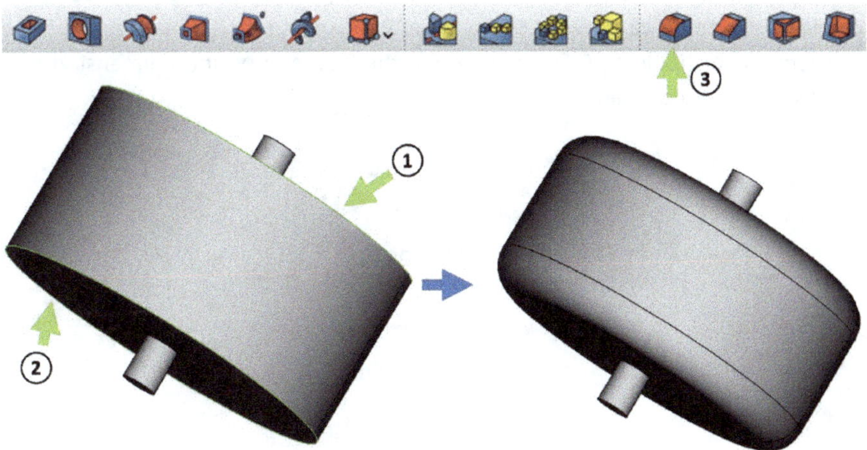

In the penultimate step, we want to improve the feel of the mouse wheel by modeling a corrugation on the surface of the mouse wheel. The best way to do this is to first hide all features and then create a new sketch on the x-y plane. In this, draw a circle with a diameter of 20 mm and its center at the coordinate origin. Additionally, transform the circle into a construction geometry. To do this, we first click on the circle ① and then on the command "Toggle construction geometry" ②. Notice the color change of the profile. We do this because the circle only serves as a reference or auxiliary structure for the profile that we are about to draw.

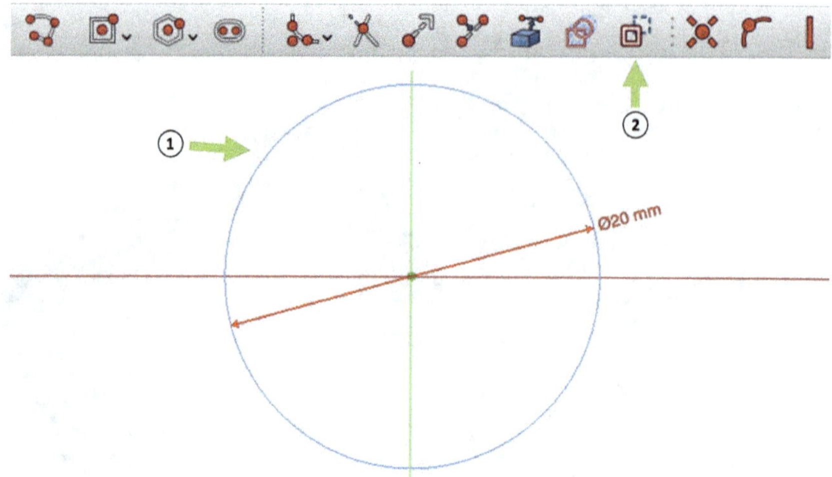

In the upper area of the circle, sketch the following profile consisting of three lines and a 3-point arc that is congruent to the circle. We can create such an arc with the command "End points and rim point" ①. To do this, we click on the two end points of the two vertical lines ② and then on the circle ③. Note: If an error occurs, you can ignore it and click it away. The vertical lines should be 0.3 mm long and the horizontal line should be 0.5 mm long. For the 0.25 mm position dimension, you must select the coordinate origin.

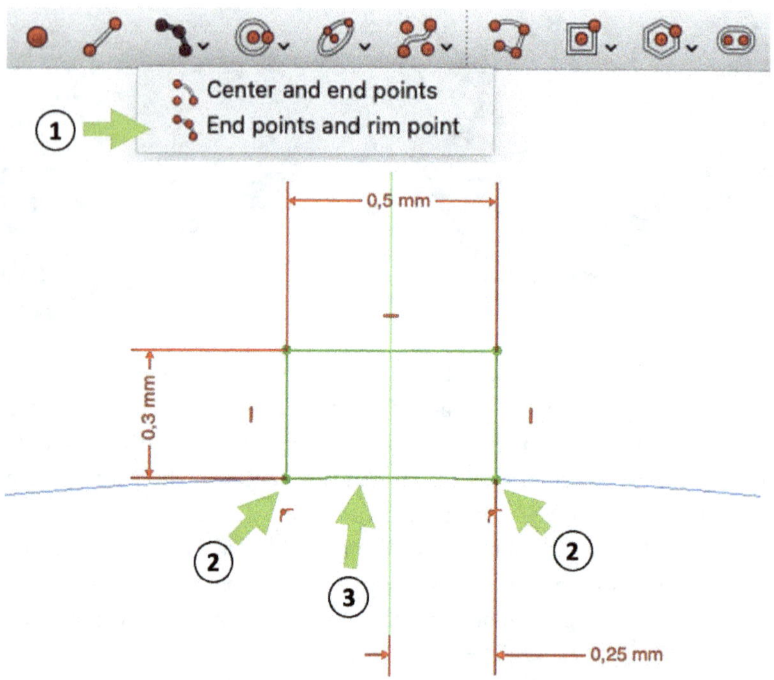

We can then close the sketch and create a two-sided extrusion with a dimension of 2.5 mm each from the 2D profile (command "Pad").

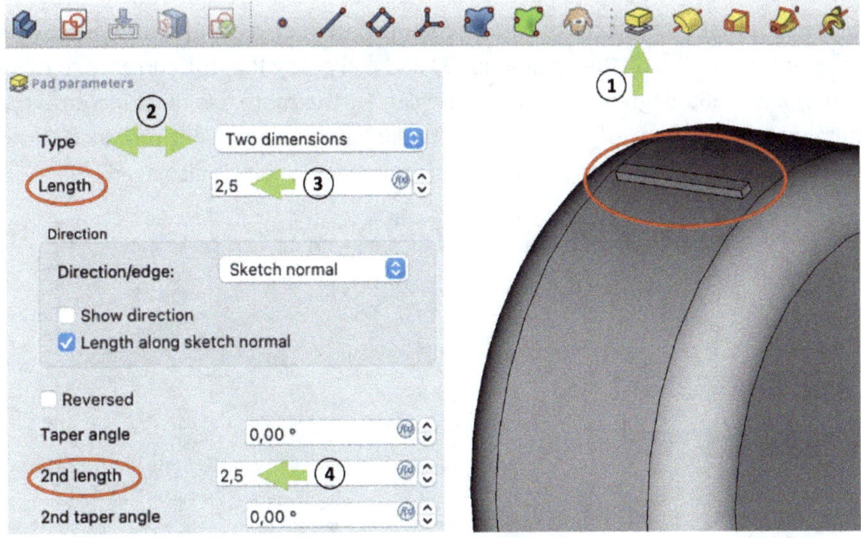

If you'd like, round off the four edges of the rectangular profile by using the command "Fillet" and selecting a radius of 0.1 mm.

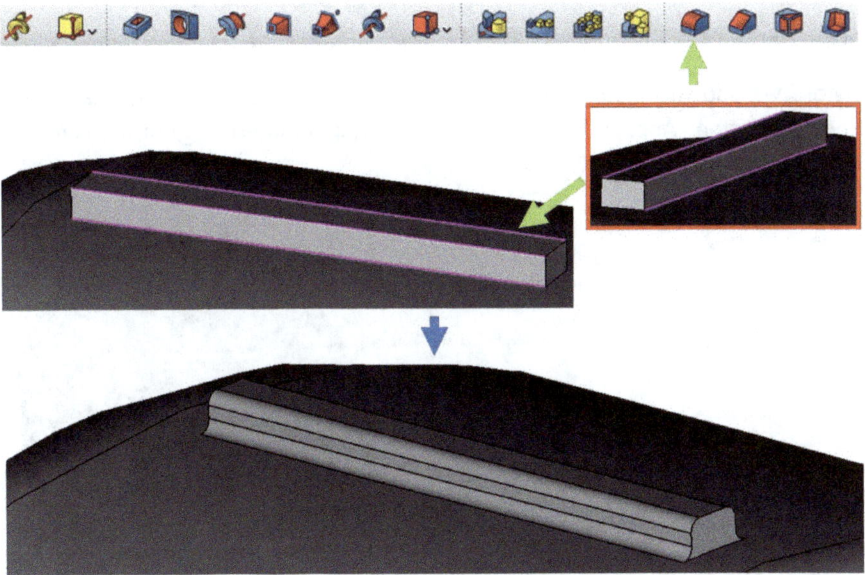

Great! How could we now create an evenly distributed fluting over the entire lateral surface of the mouse wheel from this single strut? Please think for a moment before continuing. Note: A single command is sufficient.

Exactly! We use a command to create a pattern. As we need a circular pattern here, the command "Polar Pattern" is ideal. For this command, we first select the rectangular strut ① and its fillet ② in the structure tree (hold down CTRL key). Make sure that you select the correct features, the names, and positions may be different. Then click on the command ③ and make sure that the correct axis (blue z-axis) and a 360-degree rotation ④ are set. Furthermore, we must increase the number at "Occurrences" ⑤ to 60. Depending on the system performance, it can now take a few seconds to minutes before a preview image is displayed.

Before we add the mouse wheel, we can color it. We do this with the already known function "Appearance..." ② and selecting a material ③ or a color (e.g., dark gray) ④.

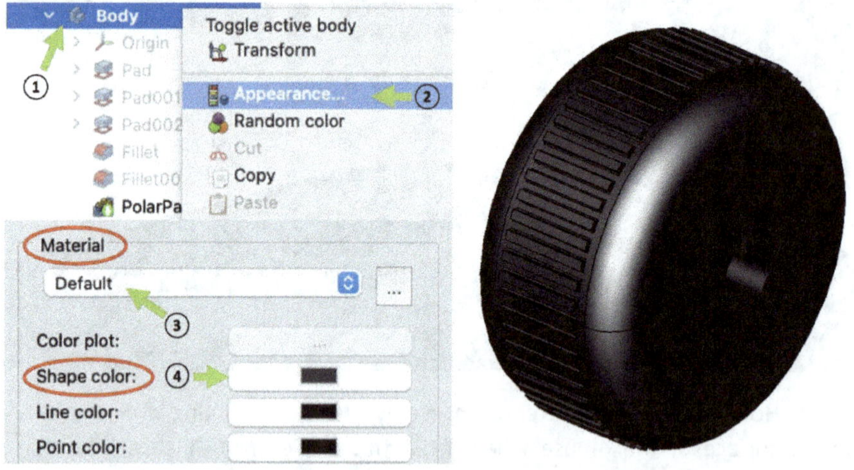

4.3 Assembly and appearance of the PC mouse

In this last chapter of the project, let's change the color of the base body first. We can do this in the same way as before. The only difference is that we cannot select a body in the combo view but have to right-click on the individual features and then select "Appearance..." and a color. You can color the PC mouse according to your taste. How about a red base body, for example, with an eye-catching golden rim?

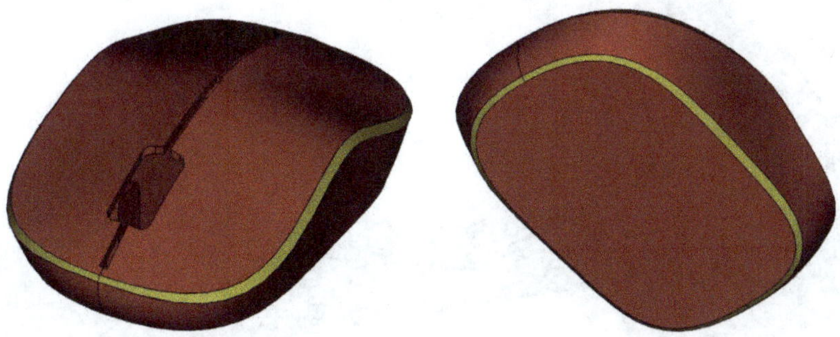

Lastly, we need to add the mouse wheel to our housing. We do this in the workspace "A2plus" ①. First add the mouse wheel to the computer mouse (command "Add a part from an external file" ②) and then use the command "Move the selected part" ③ to move and rotate it using the arrows and pivot points ④. It is best to hide the mouse buttons during this process. Excellent!

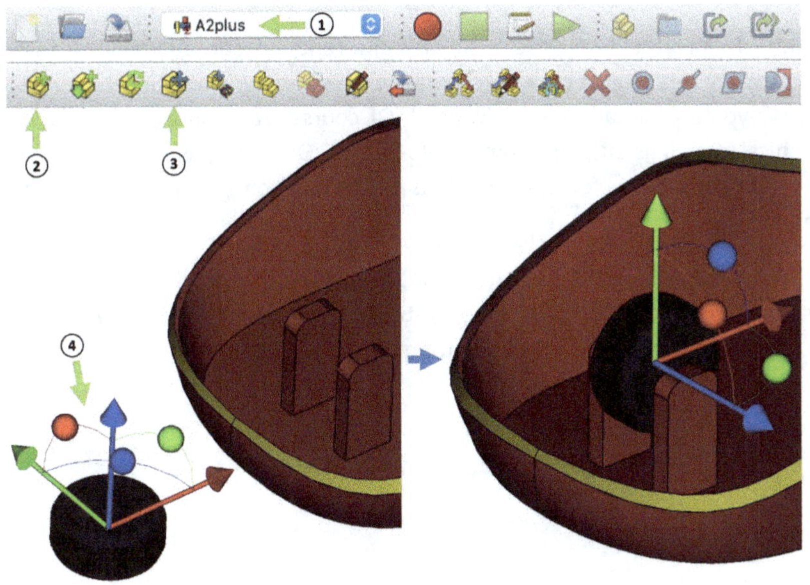

Chapter 5 | Project 4: Dumbbell bar with weight plates

Great, now you've worked your way through to the last chapter of this course! That is a strong achievement. Keep at it and be sure to complete this last project as well. You are almost there and will soon be able to call yourself a professional in "Freecad". In this chapter, we will construct the following dumbbell bar including weight plates.

In this last project, I will use a slightly different didactic approach. Please try constructing this project completely on your own. You will certainly be able to do this, as you are already very advanced. Of course, you will first receive all the technical drawings of the components below. On these, you will find all the necessary dimensions for the parts. You are also welcome to try out the assembly yourself.

Don't worry, the individual solutions for the construction of all parts including the assembly will, of course, follow step by step. If individual steps are new (e.g., fonts, knurling the handlebar), simply skip them for now.

5.1 The technical drawings of the components

10 kg weight plate:

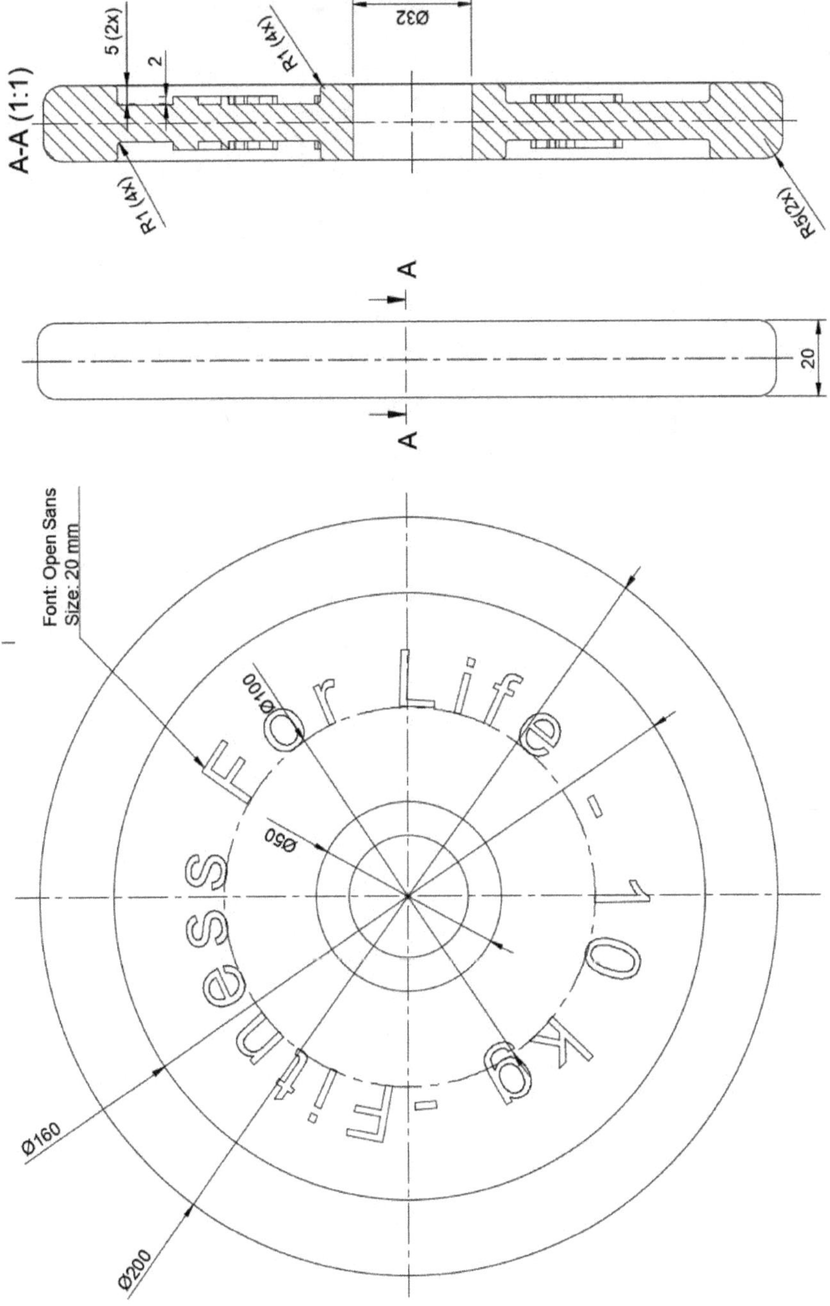

5 kg weight plate:

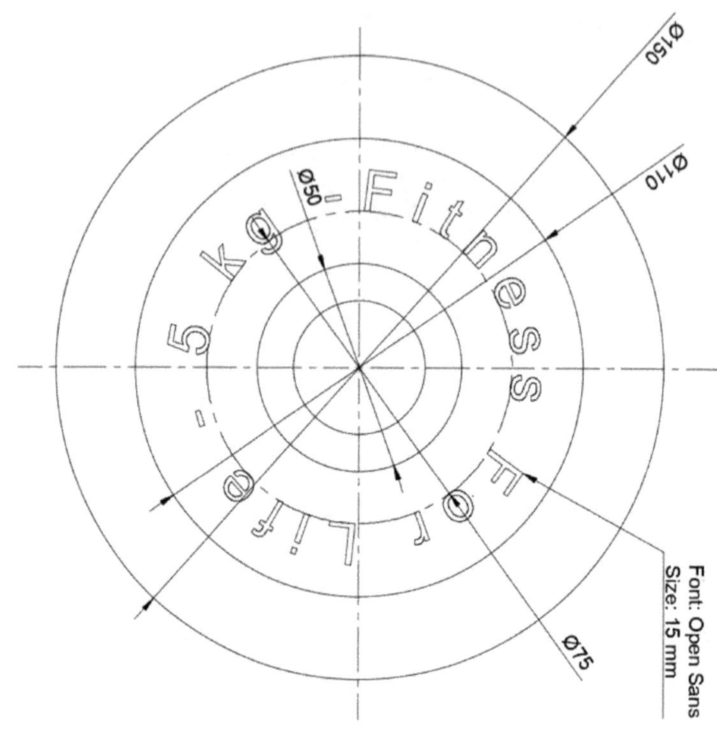

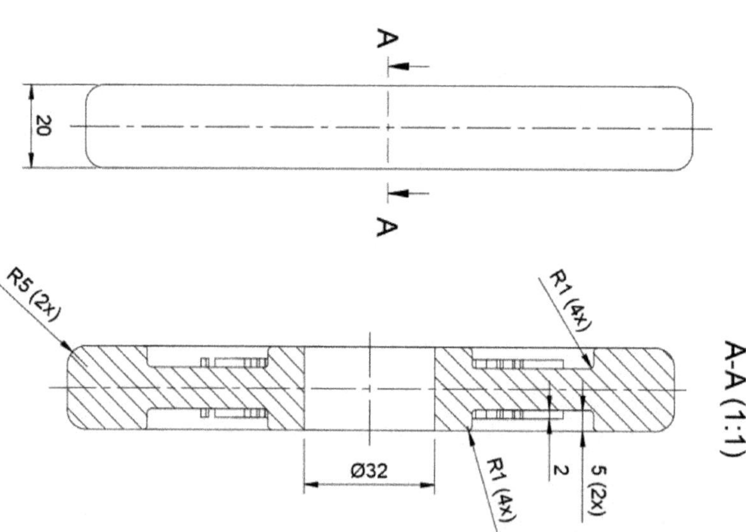

Dumbbell bar:

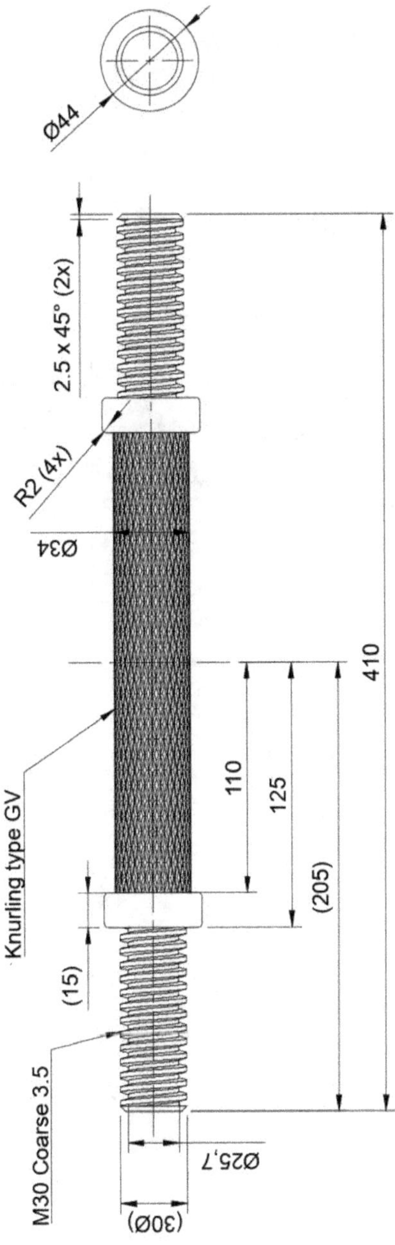

<u>Note:</u> The knurling for the handle is quite complex to create and is explained in the step-by-step section of the chapter. To simplify things, just create a smooth surface with a diameter of 34 mm first.

Star nut:

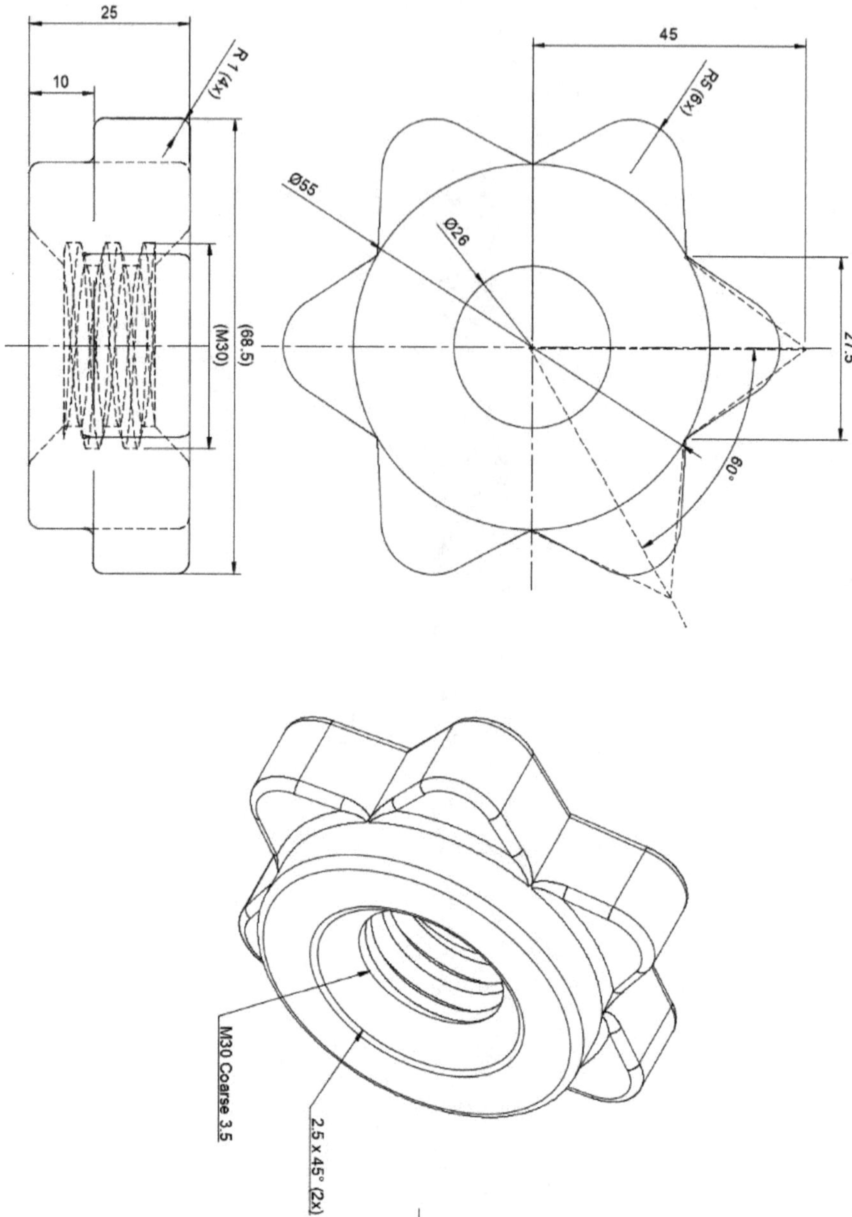

<u>Tip:</u> For the outer star geometry, first sketch the dotted triangle geometry all around and then round it off with a radius R5.

5.2 The step-by-step construction of the weight plates

For the first 10 kg weight plate, we create a new document, switch to the workspace "Part Design" ①, create a body ② and then a sketch ③ on the x-y plane. Here, we draw two circles, one with a diameter of 32 mm and one with a diameter of 200 mm. The center point should be the coordinate origin in both cases.

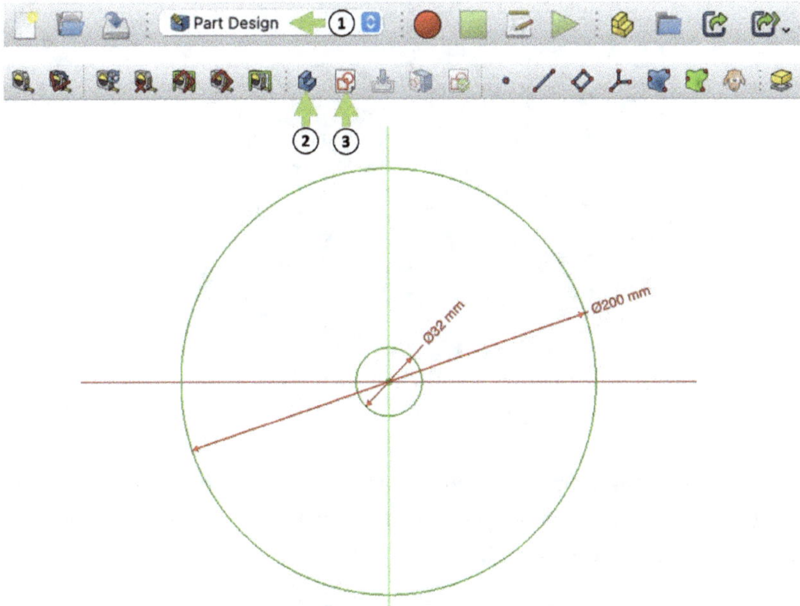

We can then finish the sketch and use the command "Pad" to create a symmetrical extrusion with a total thickness of 20 mm.

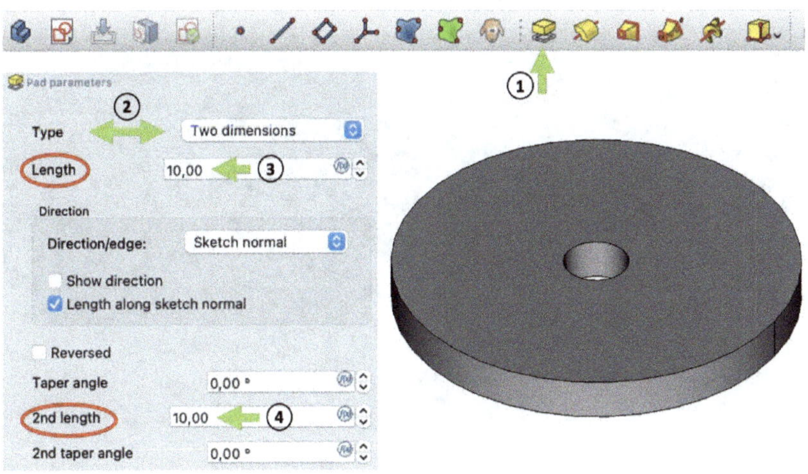

For the area of the lettering that we want to apply to the plate in the next step, we create an indentation of -5 mm on the <u>top</u> (important!) surface of the part. To do this, we need a sketch in which we draw two circles (50 mm and 160 mm diameter).

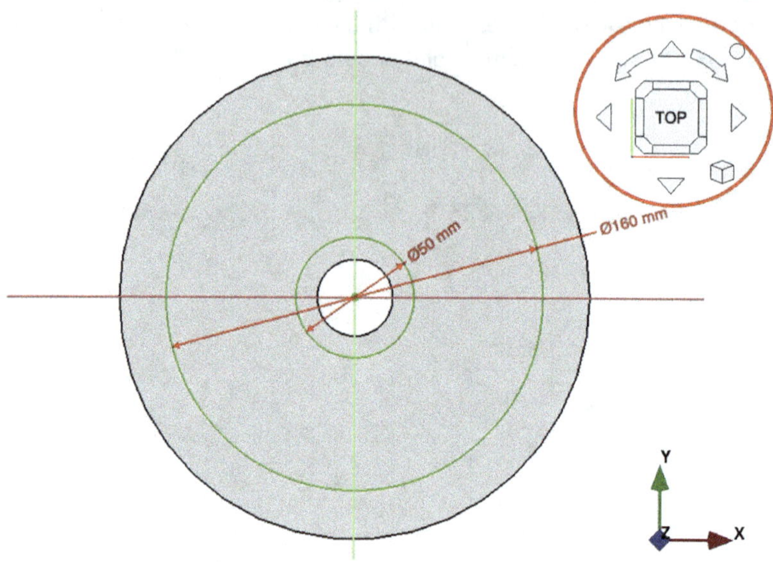

After we have finished the sketch, we create the indentation using the command "Pocket" ①. The sketch is normally recognized automatically. We then enter a value of 5 mm in the settings ②.

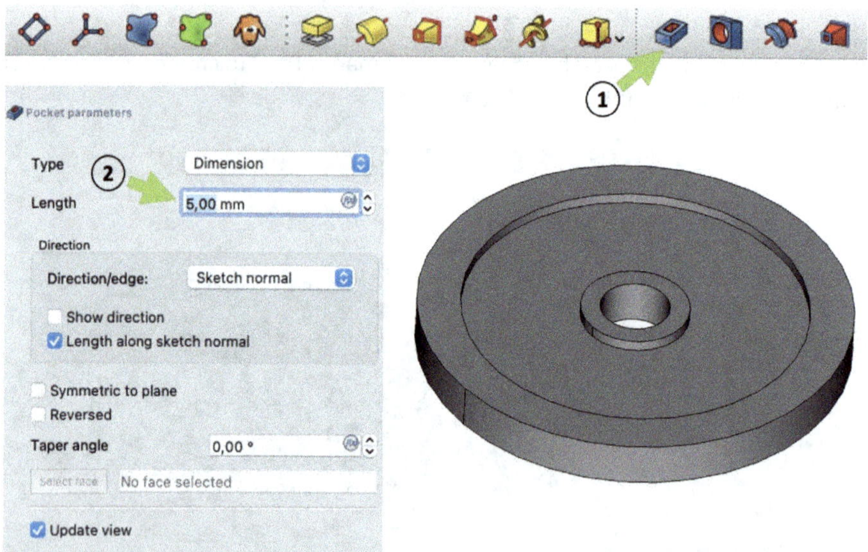

As already mentioned, we would like to add a lettering as a decorative element. We do this with the "Add-On" called "FCCircularText". This command is not installed by default. However, we can easily install the command manually using the "Add-On Manager" (① and ②).

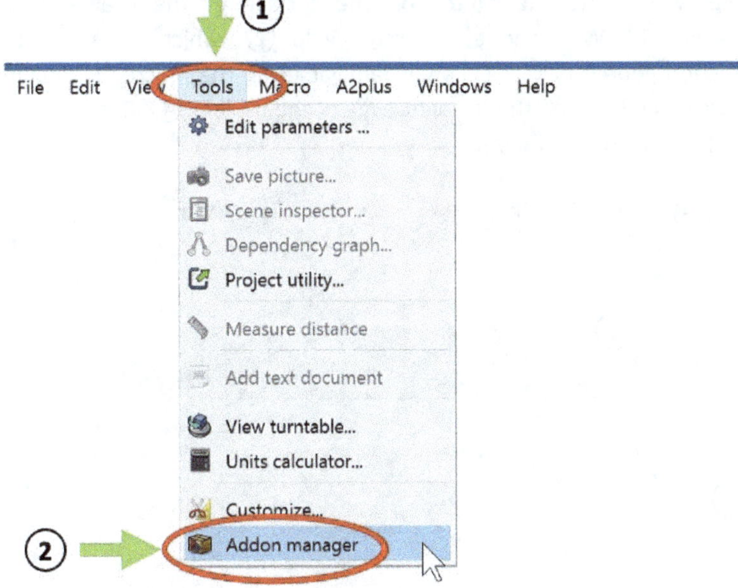

A window opens in which we have to select the search on "Macros" and then enter the abbreviation "fcc" in the search bar. We find the macro "FCCircularText" (note: there are two different ones), which we double-click on.

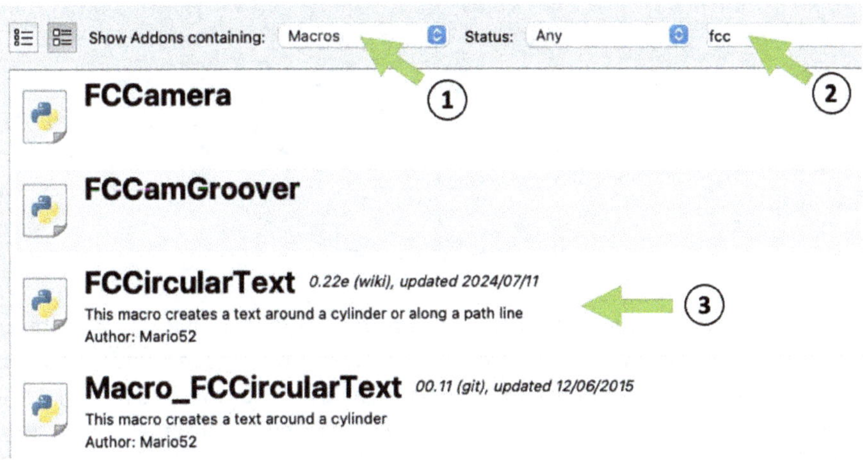

In the next window, we have to click on the button "Install" and when you are asked if you want to create a toolbar button, please click on "Yes". Then we can close the "Add-On-Manager".

Now we can start a new sketch and use the downloaded macro to create the lettering. We create the sketch on the surface of the previously created indentation ①. We need a 100 mm circle ②, which we convert into a construction geometry ③. Let's get familiar using the macro "FCCircularText", which we should find on the far-right edge of the toolbar ④. But maybe it is in a different position in your case.

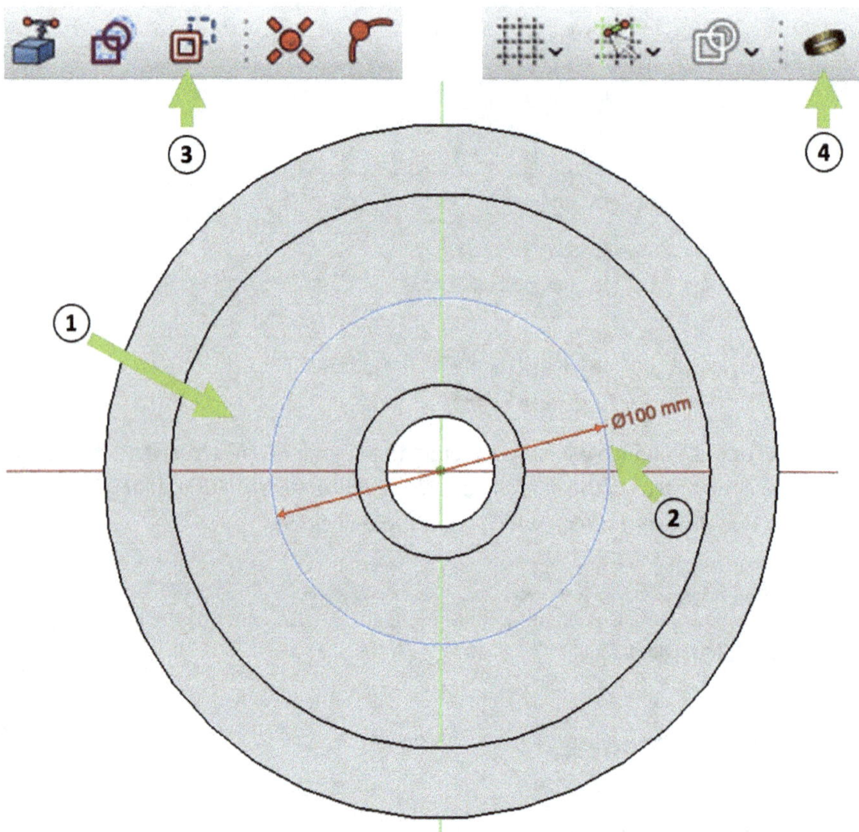

A window opens for the settings. Here we first enter the desired text, e.g., "Fitness For Life - 10 kg -" ① and in the next step we switch from "Mode Stand" to "Mode Flat" ②. Please also enter the values for the radius of the circle (70 mm) and the height of the individual letters (20 mm) ③. For the circle radius, the height of the letters must be included in this command, so the 70 mm results from 50 mm + 20 mm. As we also would like to create an extrusion at the same time, we activate

"Extrude Char." ④ and assign a value of 2 mm. The correct positioning of the font is essential, so we activate the option "Placement" ⑤. We need the coordinates of the center of the circle that we drew earlier. The coordinates are displayed in the bottom bar of "FreeCAD" when we move the mouse over the center of the circle. In this case, the values for x and y are 0 and 5.01 for z. Before we start the command in the "Command" area by clicking on "Run" ⑥, we need to download a font from the Internet and select it with "Other" ⑦ (see next step). But you could also try to see if the default font is found, and the macro runs without errors.

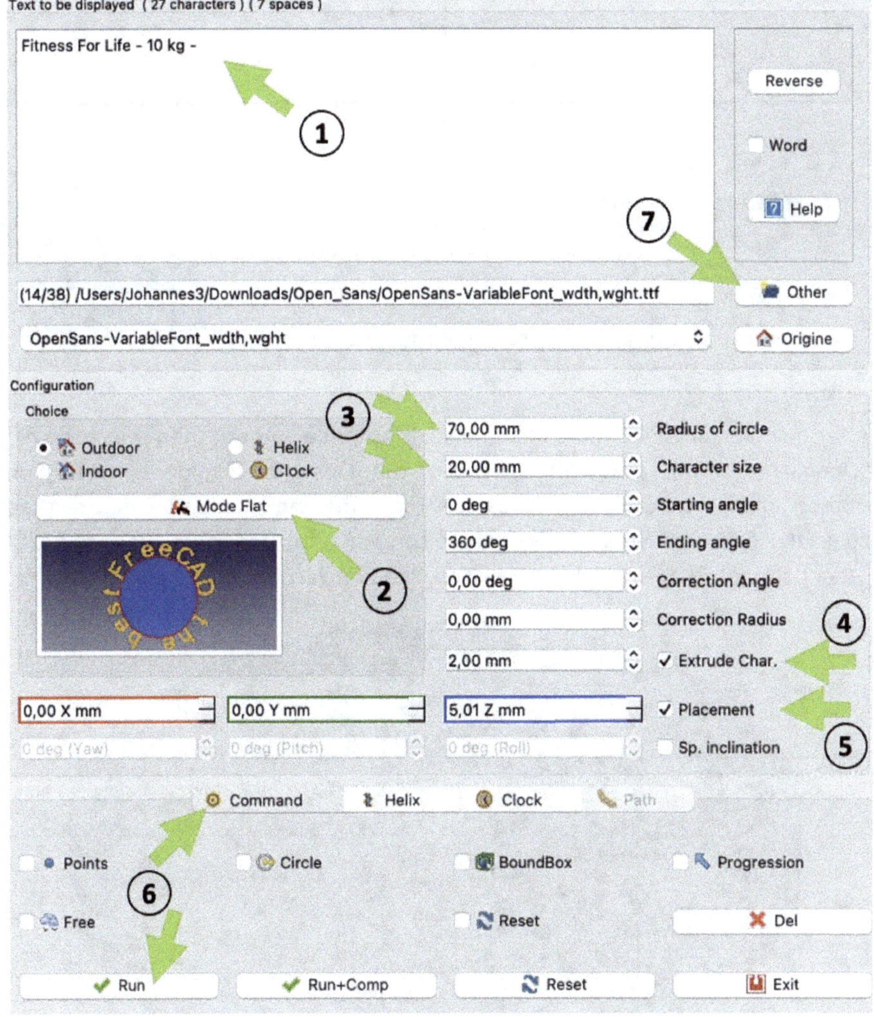

If an error occurs, it is likely that no font is found. In this case, we can download a free font from the Internet. The best way to do this is here: https://fonts.google.com/specimen/Open+Sans

First click on "Get font" and then on "Download all" in the next window.

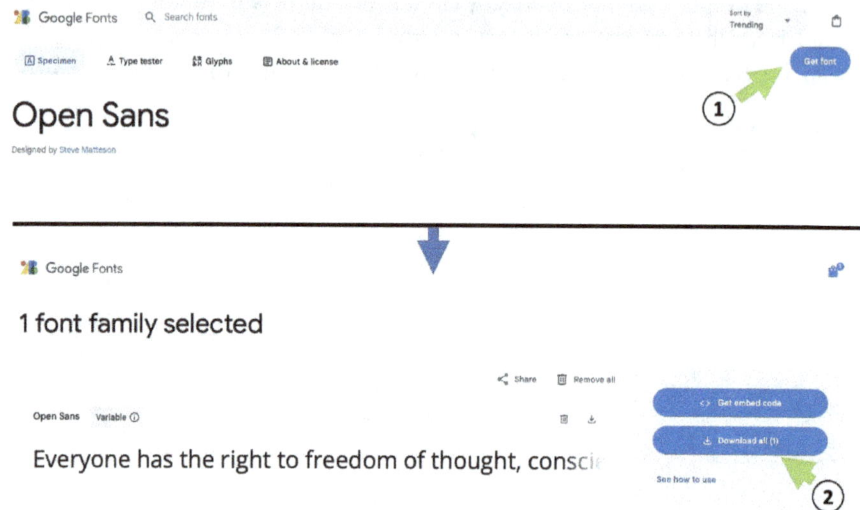

After downloading, you must first unpack the zip file and can then select the file "OpenSans-VariableFont_wdth,wght.ttf" in the "Open_Sans" folder by clicking on "Other" in the macro ⑦ (see last step). The command should now be executed correctly. If you had already executed the command for testing, you need to set all other parameters again. Then we can close the sketch and should get the following.

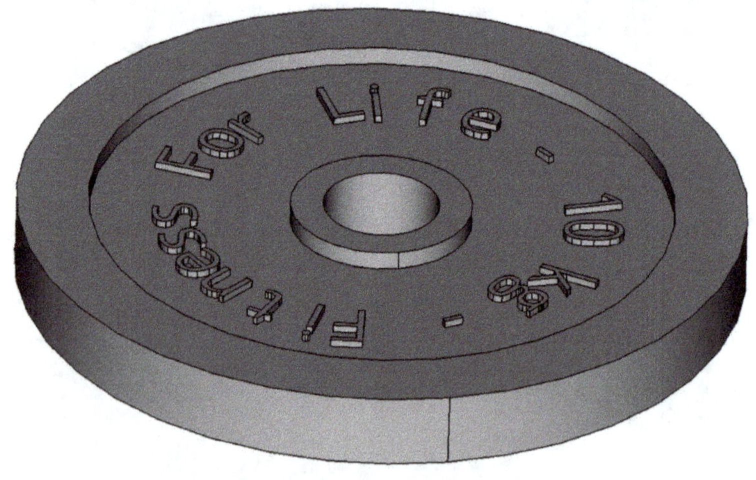

Please carry out the identical procedure on the other side. First create a 5 mm indentation, or mirror the existing one, create a sketch and then execute the macro "FCCircularText". On this side, however, we need a negative sign for the settings "Extrude Char." and "Placement", as we are moving in the negative area of the z-axis. Why? Because the x-y plane is exactly in the middle of our part.

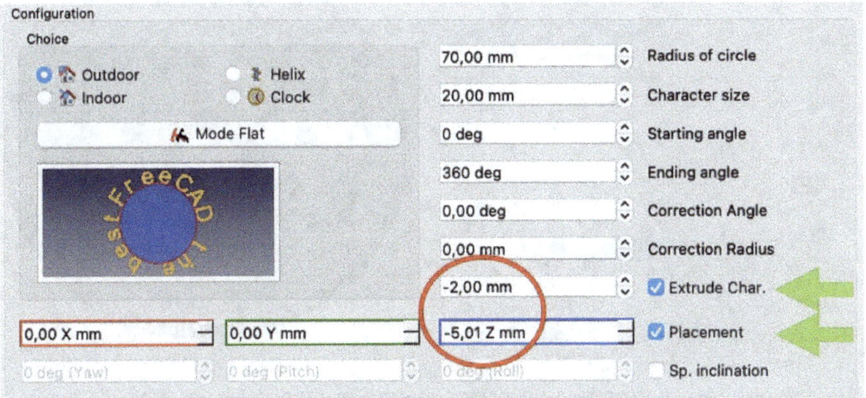

After completing the command, we can close the sketch. Finally, we can create a few fillets and color the part. For example, we round the edges ① with a radius of 5 mm and the edges ② and ③ with a radius of 1 mm. Of course, we also do this on the opposite side.

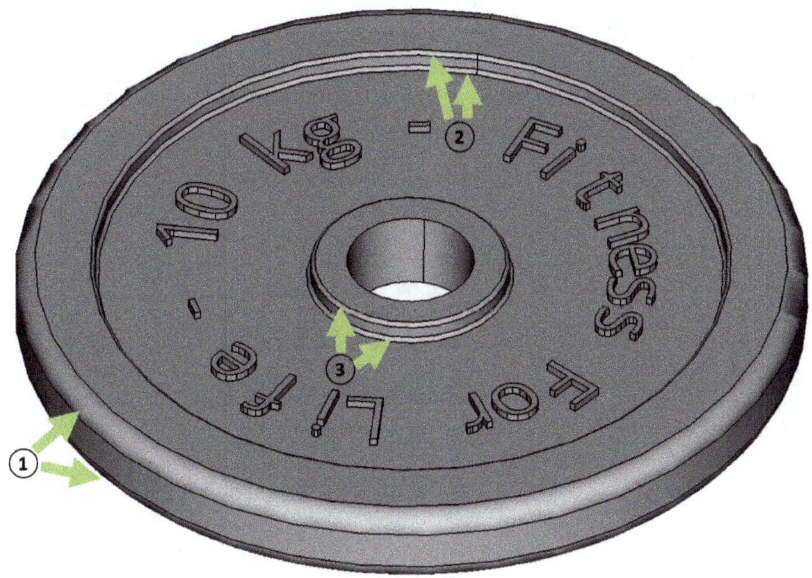

Coloring is made by using the command "Appearance..." (① and ②). For the body of the weight plate, it makes sense to select "Steel" as the material (③ and ④). However, you can of course also choose something else.

We can either leave the two letterings in gray or color them as well. To achieve this, we must first expand the two folders in which the lettering is located by clicking on the arrows ①. Then we need to select every second element (the extrusions of the individual letters) in each of the folders (hold down the CTRL key for multiple selection) and, after right-clicking, select "Appearance...". Then we can color all letters at the same time, e.g., in green.

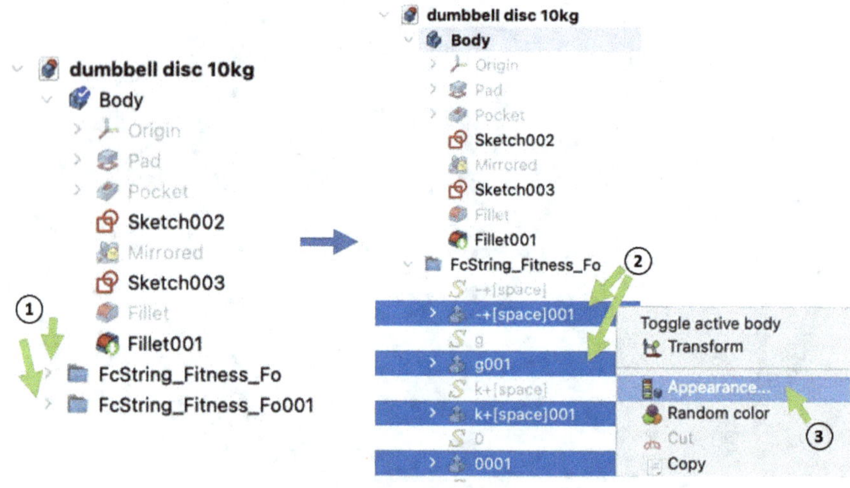

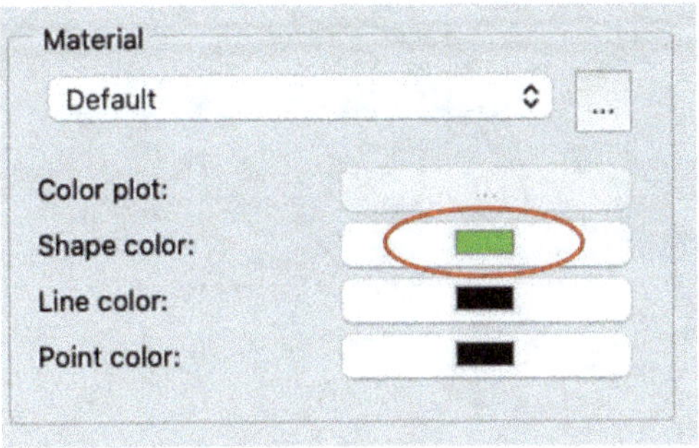

Excellent, the 10 kg weight plate is done. The 5 kg weight plate could be constructed in the same way — apart from a few dimensional differences. However, we can also simply derive the 5 kg weight plate from the 10 kg weight plate. Let's take a look at this method. It is best to save the model of the 10 kg weight plate first. Then save the same file for the 5 kg weight plate again under a different name.

First, we need to delete the two letterings, as we will make them slightly smaller later. To do this, simply select the two folders in the structure tree and delete them; you may have to confirm twice in pop-up Windows with "Yes".

The plan is to work backwards through the features in the structure tree. Therefore, we first select the "Pocket" ①, expand the folder, right-click on the corresponding sketch ②, here e.g., "Sketch001" and select the command "Edit sketch" ③. This allows us to edit the sketch. In this sketch, all we have to do is to reduce the outer circle ④ to 110 mm. We can then close the sketch. As we mirrored this feature, the indentation on the other side is automatically changed.

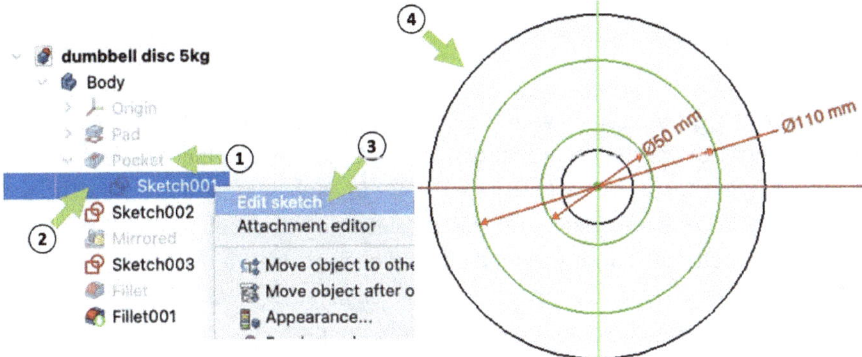

We also do this for the sketch in the "Pad" feature folder (① to ③). Here we reduce the diameter of the outer circle ④ to 150 mm.

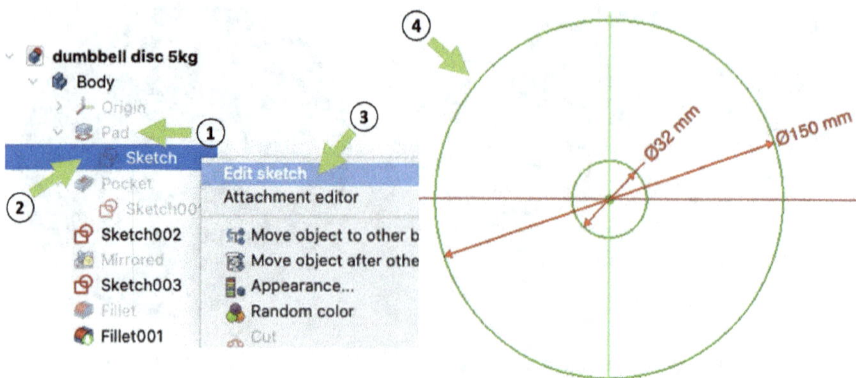

Now we create the new lettering. To do this, we first edit the sketches (① and ②) containing the construction circles in the same way as before. We change the diameter of the construction circles ③ from 100 mm to 75 mm.

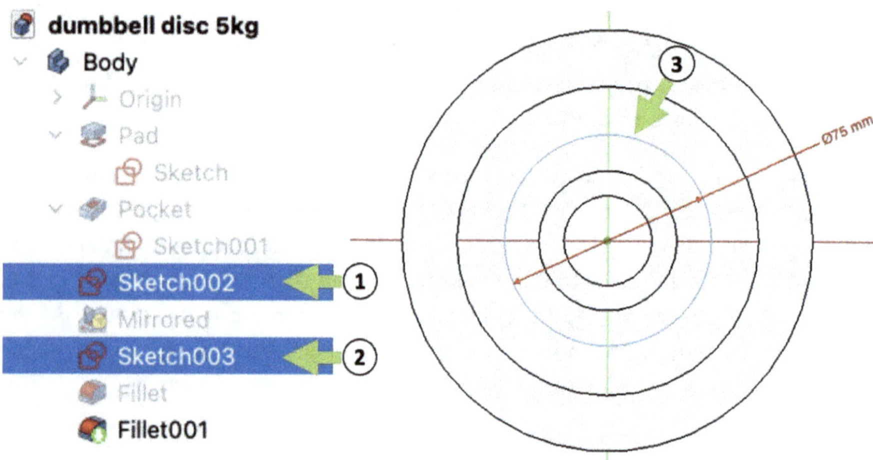

We make the lettering again with the macro "FCCircularText". The settings for the command are almost identical to those for the 10 kg weight plate. We just need to use a different text "Fitness For Life - 5kg -" ① as well as a different radius and font size. The required radius ③ for the lettering is calculated by halving the desired diameter of 75 mm and then adding the font size, i.e., 37.5 mm + 15 mm = 52.5 mm. Please remember to set negative signs at ④ and ⑤ for the second lettering, which is located on the underside of the part.

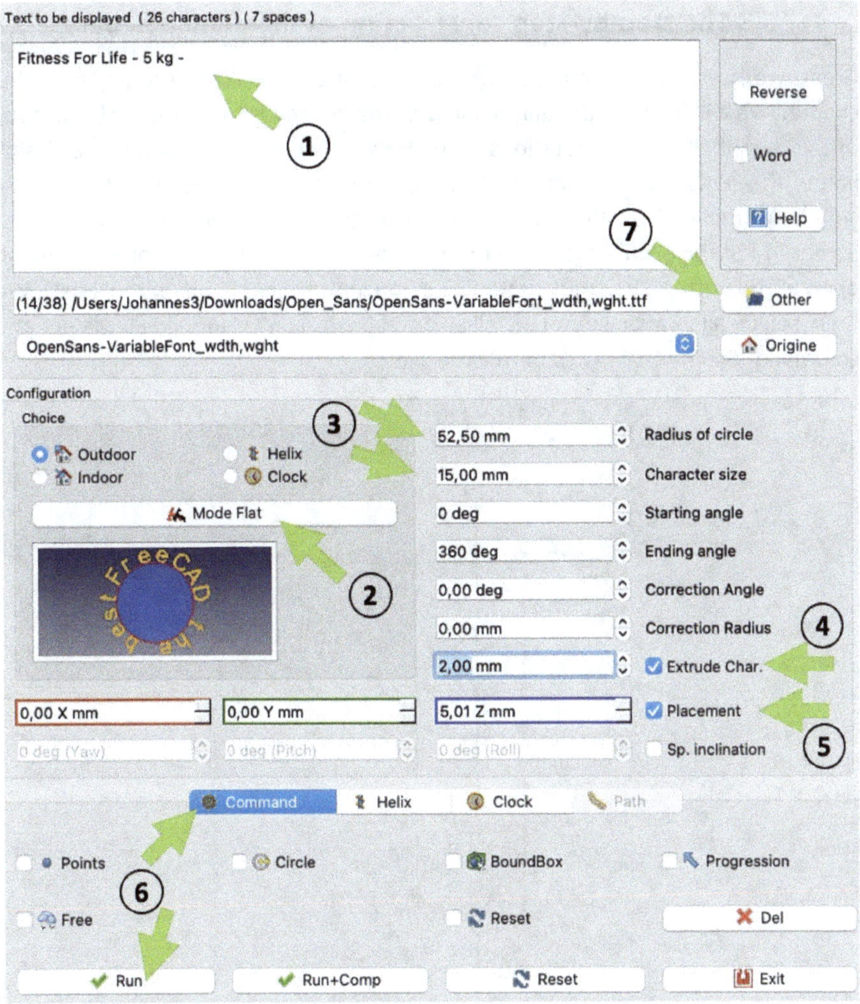

Text to be displayed (26 characters) (7 spaces)

Fitness For Life - 5 kg -

1

Reverse

Word

? Help

7

(14/38) /Users/Johannes3/Downloads/Open_Sans/OpenSans-VariableFont_wdth,wght.ttf

Other

OpenSans-VariableFont_wdth,wght

Origine

Configuration

Choice

- Outdoor
- Indoor
- Helix
- Clock

Mode Flat

2

3

52,50 mm	Radius of circle
15,00 mm	Character size
0 deg	Starting angle
360 deg	Ending angle
0,00 deg	Correction Angle
0,00 mm	Correction Radius
2,00 mm	☑ Extrude Char.

4

0,00 X mm   0,00 Y mm   5,01 Z mm   ☑ Placement

0 deg (Yaw)   0 deg (Pitch)   0 deg (Roll)   ☐ Sp. inclination

5

Command	Helix	Clock	Path

- ☐ Points
- ☐ Free
- ☐ Circle
- ☐ BoundBox
- ☐ Reset
- ☐ Progression
- ✗ Del

6

✓ Run   ✓ Run+Comp   Reset   Exit

Finally, we can also color the lettering of this part.

5.3 The step-by-step construction of the dumbbell bar

Great, we have now already constructed two of the four required parts. In this chapter, we will focus on the dumbbell bar. It makes sense to create the basic body of the dumbbell bar as a rotational part. How do we do this? Exactly, we sketch half of the cross-section of one half of the bar on the x-z **plane** (important!). We will then simply mirror this half of the bar to get the full part. The bottom-right corner ① of the profile should lie on the coordinate origin. The profile should be open at the bottom, i.e., must <u>not</u> have a horizontal connection on the x-axis. We then select <u>all</u> drawn lines ②, click on the x-axis ③ and finally select the command "Symmetry" ④ to mirror the sketch.

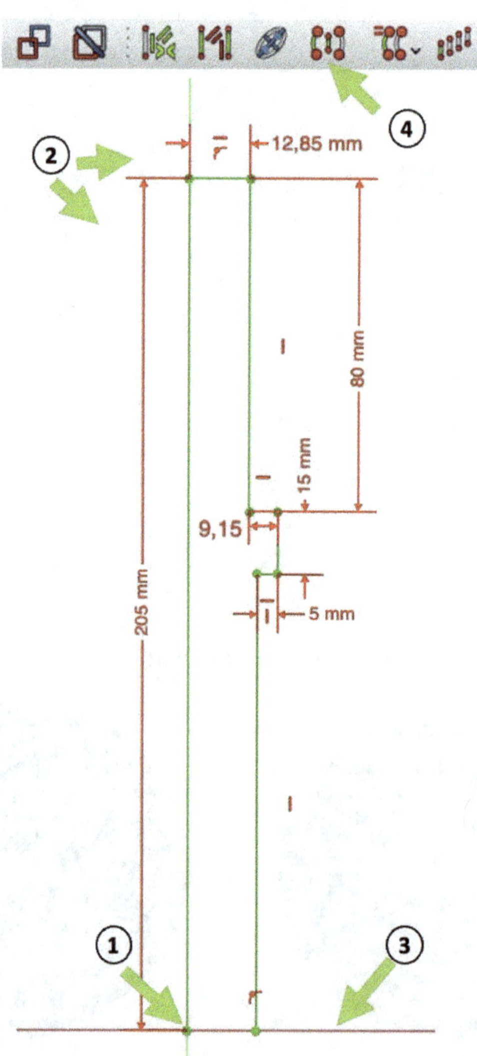

After closing the sketch, we should have obtained the following 2D profile (1). We can rotate it by 360 degrees into a 3D part using the command "Revolution" (2). Additionally, we create a 2.5 mm chamfer at each end of the bar ((3) - (5)).

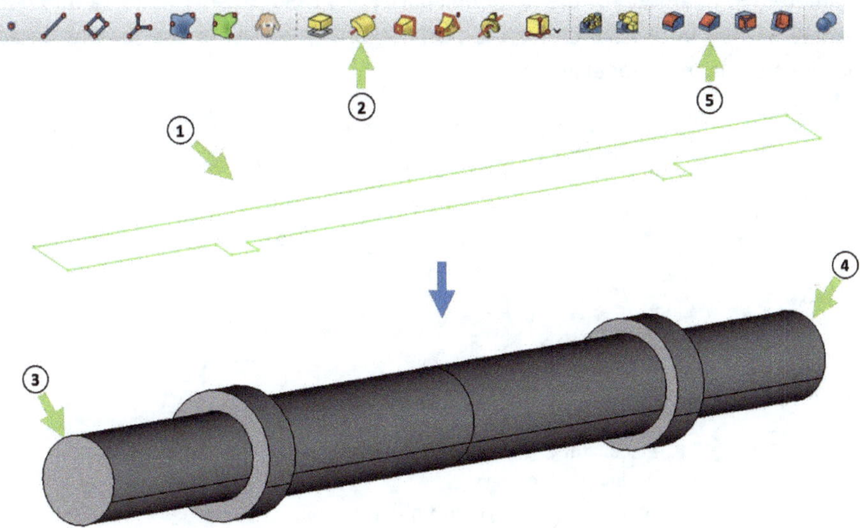

Now we create threads at the ends of the dumbbell bar. To do this, we need the "ThreadProfile" workspace, which is <u>not</u> installed by default. If you have not already installed it for a previous project, you can install it — similar to the macro "FCCircularText" — via the "Add-On Manager". To achieve this, search in the "Workbenches" section (1) for "thread" (2).

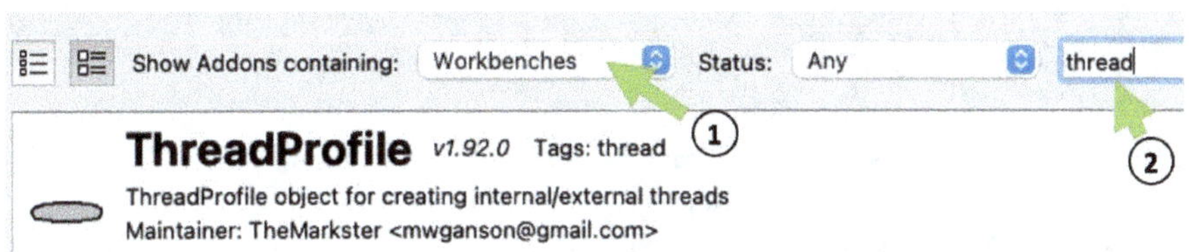

To create the first thread, we can then switch to the "ThreadProfile" (1) workspace that has just been installed. A thread always consists of a profile, a helix and a 3D feature ("Sweep").

In the first step, we create the thread profile by clicking on the button "Create V thread profile" (2). This allows us to make settings in the "Data" (3) section of the combination view. For correct positioning, we expand the folders "Placement" (4) and "Position" (5) and enter a value of -202 mm for "z" (6). This value is the negative end of the dumbbell bar (205 mm length in positive and negative z-axis direction minus 2.5 mm chamfer minus 0.5 mm as additional initial offset). For the option "Height" ⑦ we enter the desired length of the thread, the value should be 76 mm. Finally, we select the desired thread profile at "Presets" ⑧. We need a "M30 Coarse 3.5" thread. At "Minor Diameter" we can read the required core diameter ⑨ (diameter of the dumbbell bar). For a M30 thread it is 25.71 mm (we had 12.85 mm in our sketch for the profile <u>half</u>). This value must be considered before creating the model and a thread.

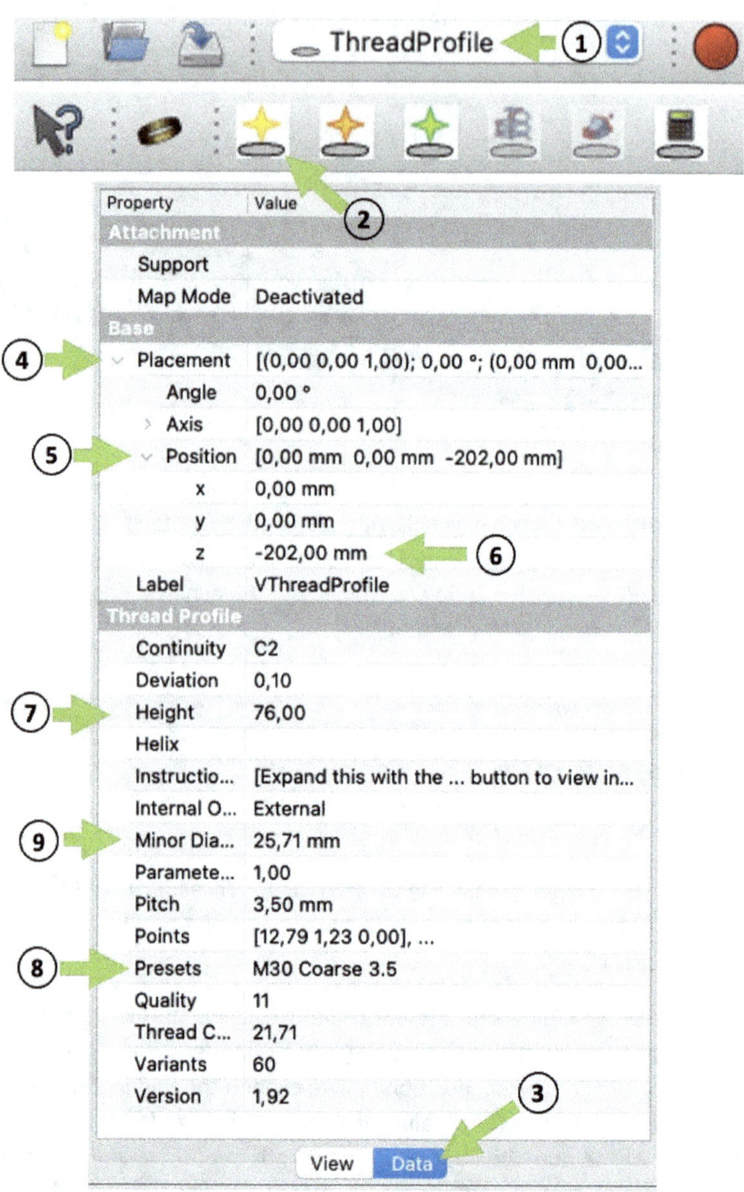

To make the helix and the 3D feature, we only need to click on the two commands "Make Helix" ① and "Do Sweep" ② one after the other. Make sure that you execute these commands immediately after the first command so that the correct features are still selected in the structure tree. Otherwise, you may have to select them again. Depending on the processing power of your PC, the thread creation can take a few seconds to minutes. If the process works, you will receive the desired thread.

We must repeat this procedure on the other side of the barbell. This works in the same way as before, the only difference is that **+202** mm must be entered for the position variable "z" (⑥ in the previous step), as the thread expands in positive axis direction. All other values can be taken from the previous step.

Now we have almost finished the dumbbell bar. In the last step, we can create a knurling on the grip surface to improve the feel. Please save the previous part first! Creating the knurling is very computationally intensive and may cause the program to crash. We create the knurling by generating a polar pattern of two opposing subtractive helices (plural of helix). This sounds complicated, so let's take a look at each step. First, we need a sketch, which we create on the inner left attachment

(1) of the rod. Important: Make sure that the component is rotated in a way, so that the blue z-axis of the coordinate system points in the positive direction — i.e., to the right.

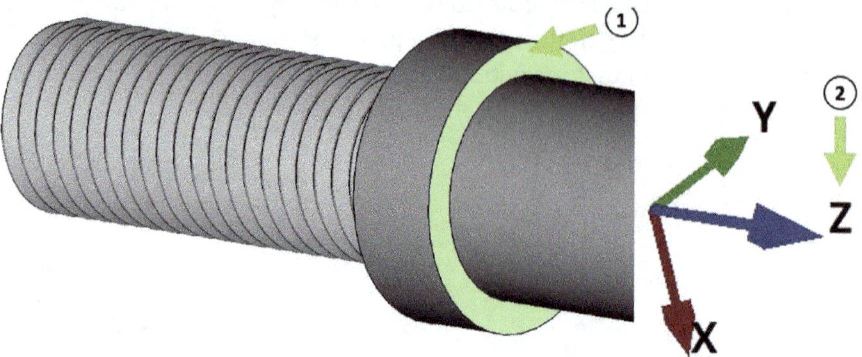

Before we sketch, it is best to hide all previous features first (space bar). In the sketch, we draw the following triangular profile, which is made up of two lines and a 3-point arc (command "End points and rim points"). The profile should be located at the upper end of a 35 mm circle. The center of the circle should be at the origin of the coordinates. If an error appears when creating the 3-point arc, you can ignore it. Important: We must then convert the 35 mm circle into a construction geometry (command: "Toggle construction geometry"). Notice the color change.

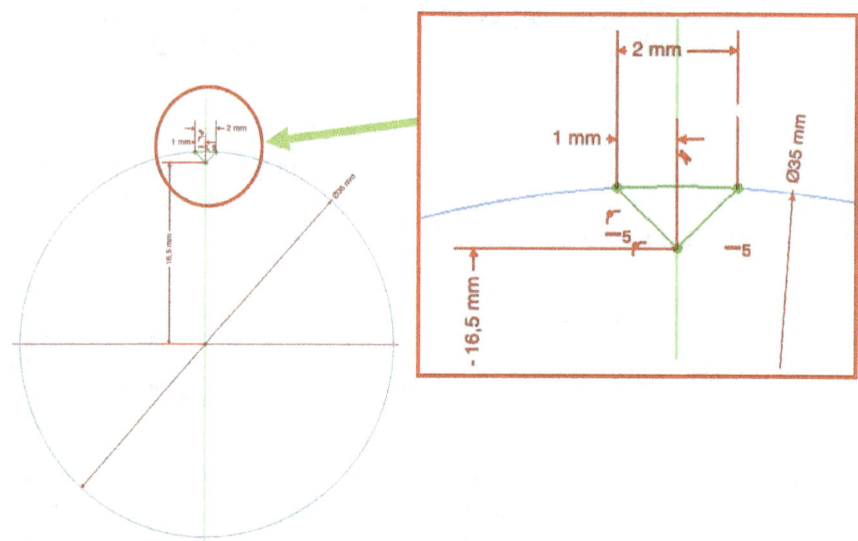

We can then close the sketch and create the first subtractive helix. We do this by selecting the sketch we have just created (1) and clicking on the command

"Subtractive Helix" ②. An error message will most likely appear, which you can ignore. We first need to adjust the settings for the command to work. We therefore select the correct axis "Base Z axis" ③, then the dimensioning mode "Height-Turns-Angle" ④ and a height of 220 mm ⑤ (length of the inside of the bar) as well as 0.5 as the number of turns ⑥. Confirm with "OK".

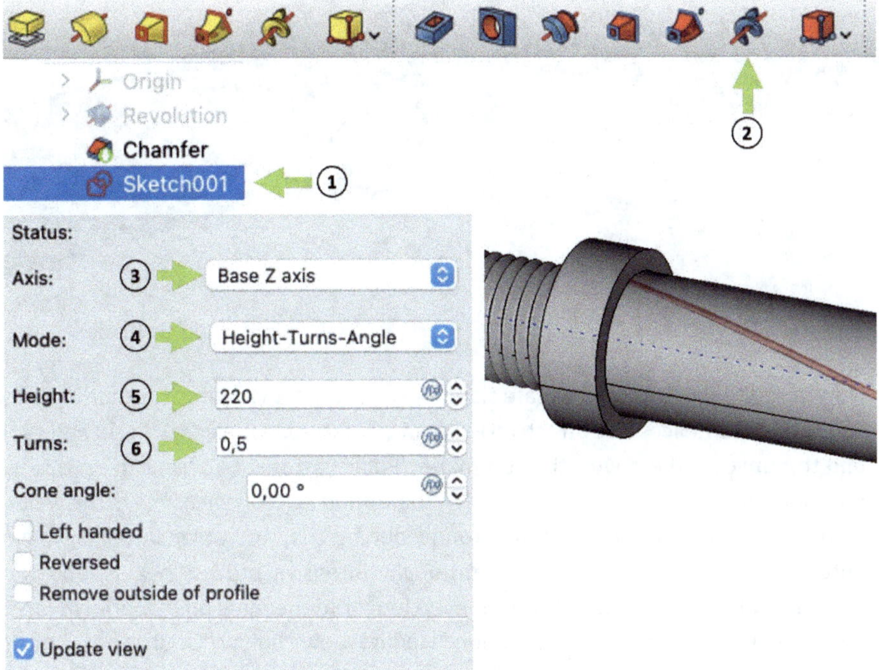

Then we do this procedure one more time. We can use the same sketch, which was moved to the feature "Subtractive Helix" ①. You will probably need to expand the feature by clicking on the small arrow.

The command is executed in the same way as before. The only difference to the previous step is that we additionally activate the option "Left handed" ⑦, in order to reverse the alignment of the second helix. Then we can confirm with "OK".

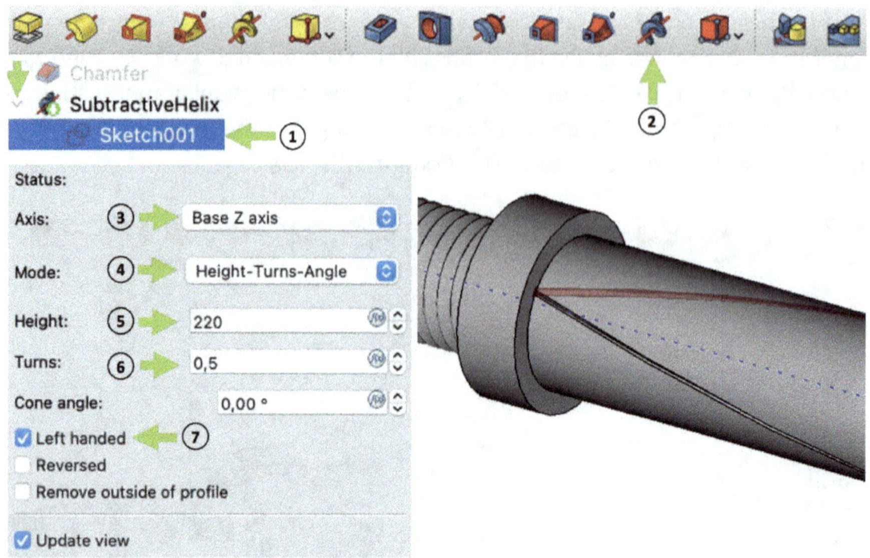

Now save the file to be on the safe side. The last step for creating the knurling is to select the two helices ("SubtractiveHelix" and "SubtractiveHelix001") (① and ②) and to duplicate them with the command "Polar Pattern" ③. We can do this by entering the value 30 in the settings for the option "Occurrences" ④. This value results in a nice structure, which I found out by trial and error. After we have entered the value, the very computationally intensive process begins, and the program will run for 5 - 10 minutes or even longer (depending on PC performance) and probably stops responding (option "Update view" below "Occurrences" must be active). Now is the best time to make a sandwich or a coffee, just check the progress in a few minutes.

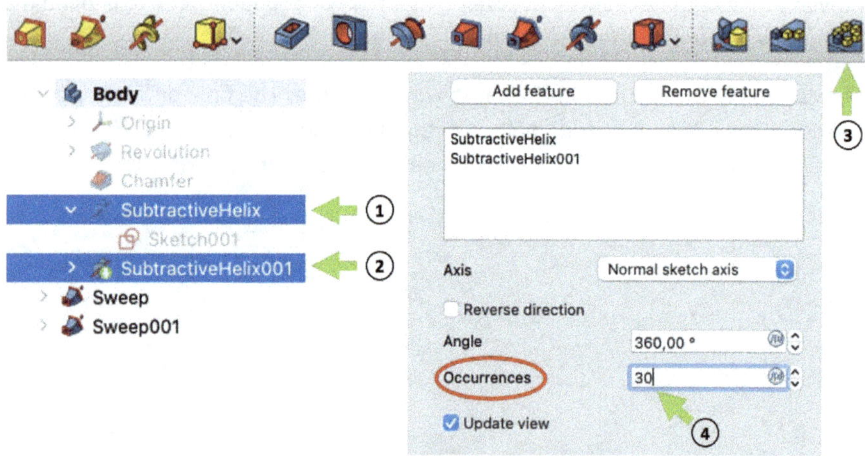

As soon as the program has completed the process, we should be able to see an even knurl structure. However, this is only the preview. We have to finally execute the command with "OK".

Great, the most difficult steps have been mastered! Now we can round the following four edges with a radius of 2 mm each and select a material or color for the part (command "Appearances...").

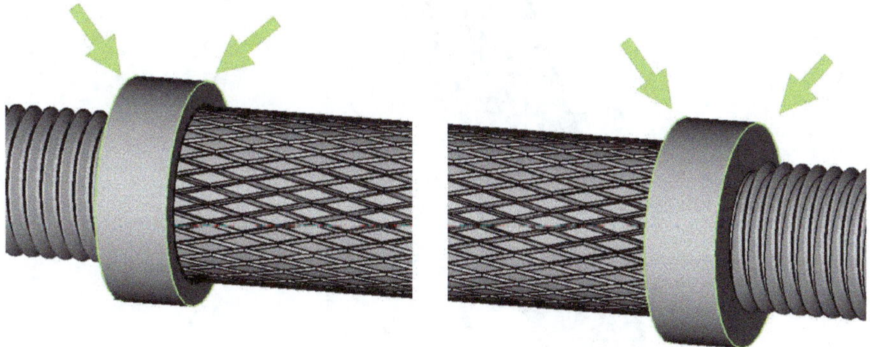

In this case, how about the material "Chrome"? You are also welcome to select a different material or maybe a fancy color.

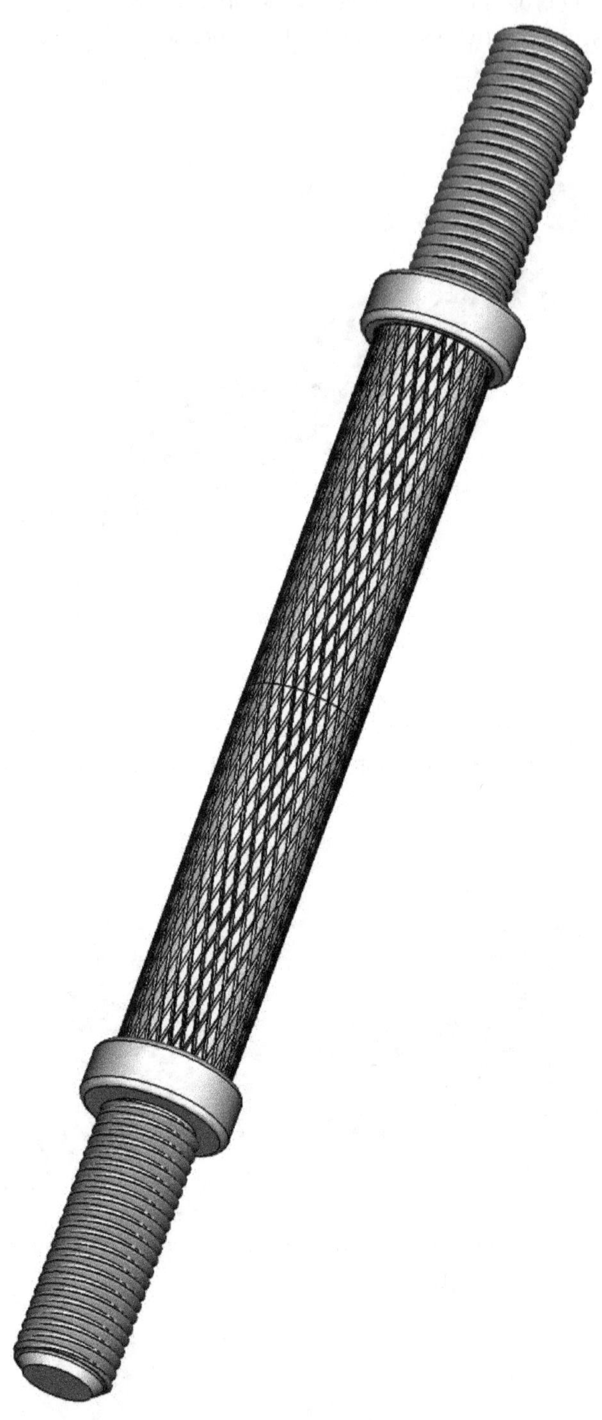

5.4 The step-by-step construction of the star nut

The last part that we will construct for our dumbbell set is the star nut. Two of them are screwed onto the ends of the bars to hold the weight plates in place. We create a body for this star nut in a new document and then start a sketch on the x-y plane first. In this sketch, we draw a 26 mm and a 55 mm circle, whose starting points should each lie at the origin of the coordinates. This will be the basic solid.

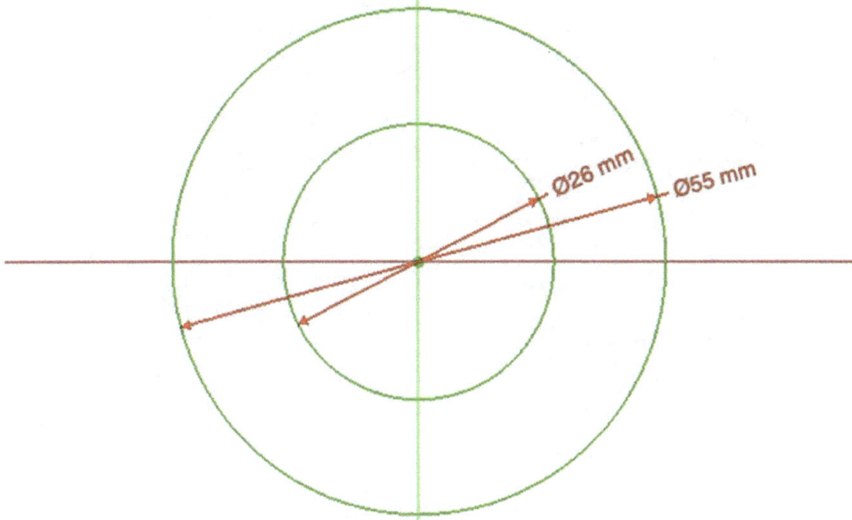

With this sketch and the command "Pad" we create a 10 mm long 3D body (only in positive z-axis direction; no symmetrical extrusion). We then create another sketch on the x-y plane in which we want to draw the star-shaped part of the component.

To do this, we hide the previous component and draw another 26 mm and a 55 mm circle ①. We convert the larger circle into a construction geometry and sketch two lines in the upper area to create a triangle that is open at the bottom ②. The end points should lie on the green y-axis and on the 55 mm circle. We then draw two diagonal lines that connect the lower end points of the triangle with the coordinate origin and convert these into construction geometries ③.

Next, we use the command "Symmetry" three times to create the remaining points of the star. It is important to select the lines to be mirrored first and to click on the symmetry line last before clicking on the command. The symmetry lines are the two diagonal lines ③ for the first two mirrorings and the red x-axis ④ in the last mirroring.

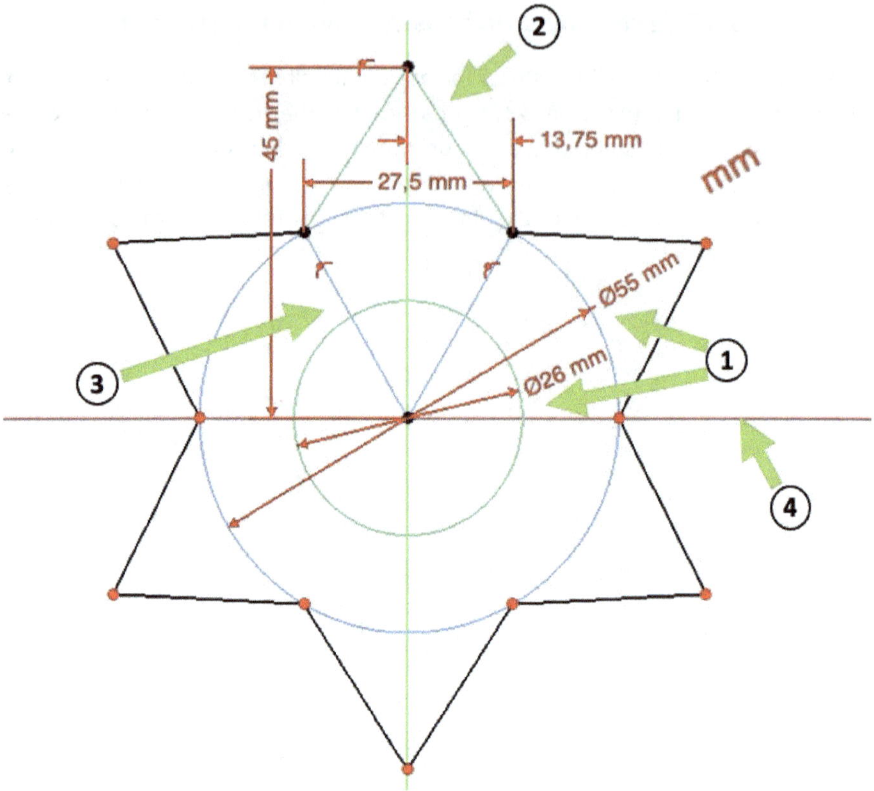

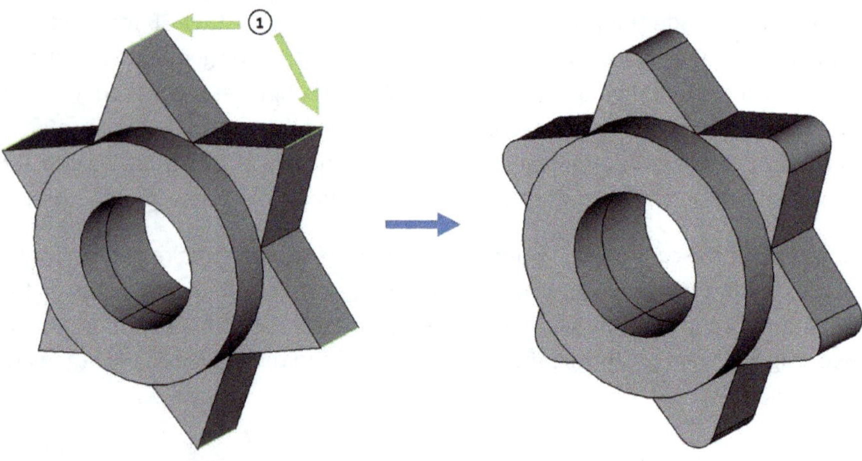

We can then close the sketch and create a 15 mm extrusion in the negative z-axis direction (command "Pad"). We also round the six outer edges ① of the star geometry with a radius of 5 mm each (command "Fillet").

Before we create the internal thread, we want to add two 2.5 mm chamfers on the two outer edges ① of the hole (command "Chamfer").

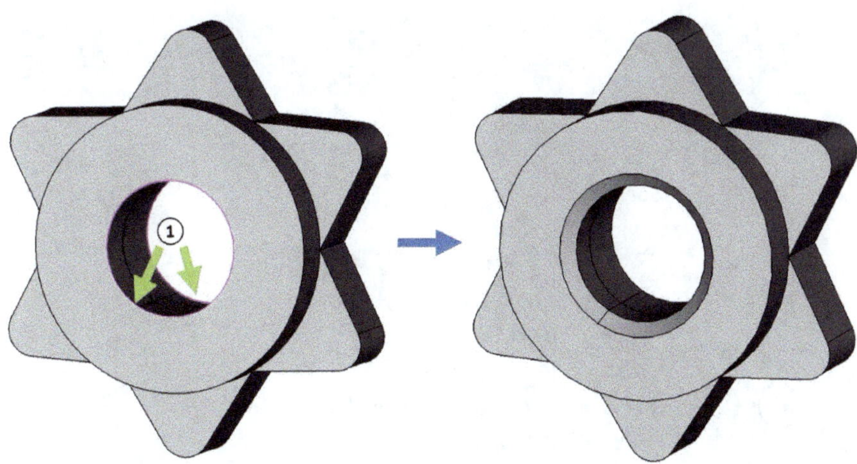

Now we can switch to the workspace "ThreadProfile" ①, in order to create the internal thread. This process works in the same way as before for the external thread of the barbell. Only the parameters are slightly different.

In the first step, we create a thread profile by clicking on the button "Create V thread profile" ②. To specify the positioning, we expand the folders "Placement" ③ and "Position" ④ and enter a value of -12.5 mm in the "z" field ⑤. This value is the negative end of the inner hole (we need to move 15 mm in the negative z-axis direction minus 2.5 mm chamfer starting in the coordinate origin).

We enter the length of the thread at "Height" ⑥. We need 20 mm, as the part is 25 mm deep in total, and we need to subtract 2 × 2.5 mm for the chamfers. To create an internal thread, we change the setting "InternalOrExternal" to "Internal" ⑦. Then we also have to select the thread profile "M30 Coarse 3.5" ⑧ and can then read off the required internal diameter of 26.21 mm ⑨.

In our previous sketches, I rounded this diameter to 26 mm. A smaller diameter will also work with this command for the same internal thread, but a larger one will not. You are welcome to try this on your own.

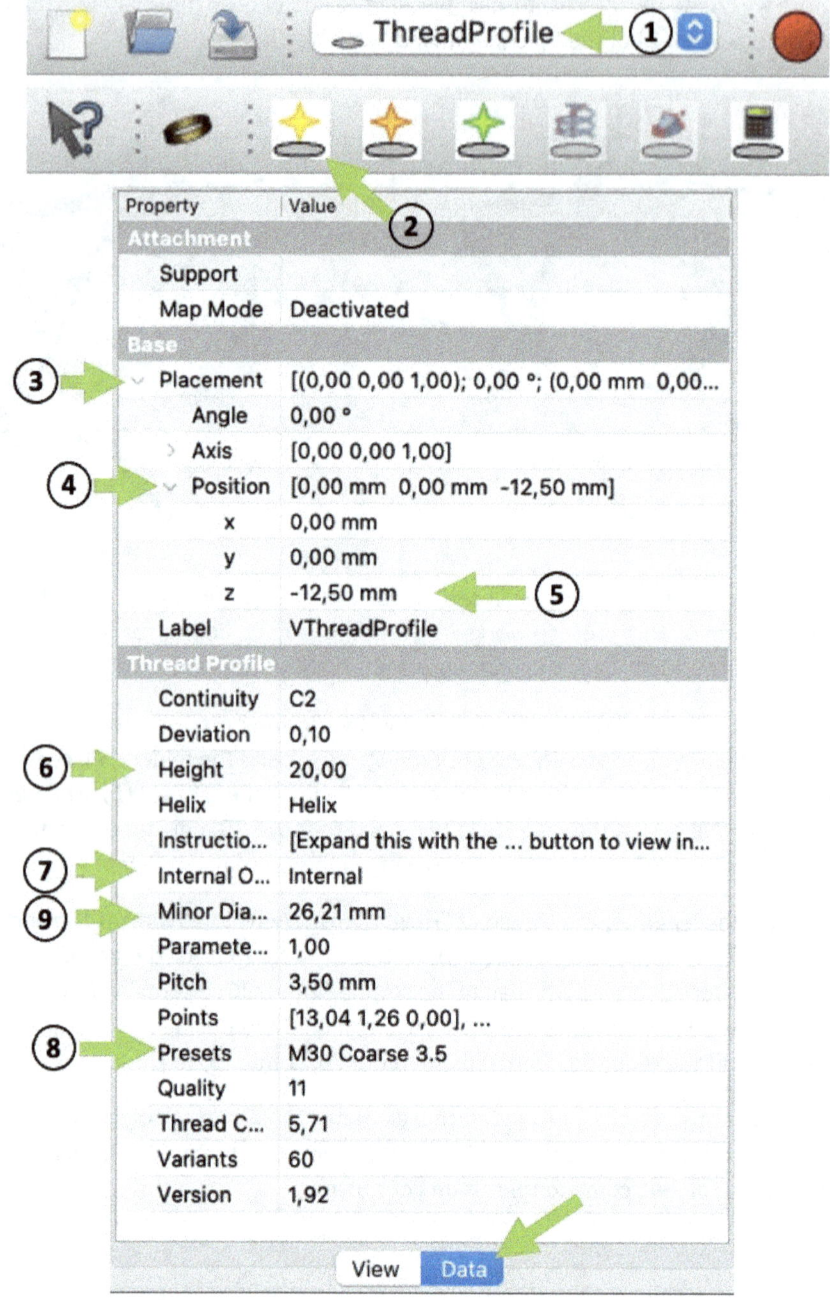

Immediately afterward, we click on the two commands "Make Helix" ① and "Do Sweep" ② and confirm the second command with "OK". This creates the internal thread.

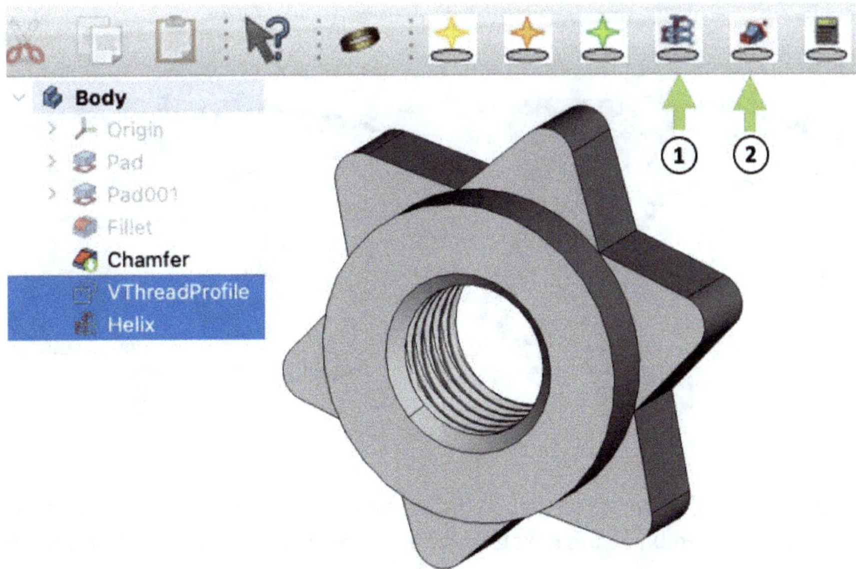

In the last two steps, we round off a few edges and then assign a material to the part. For the edges, we simply select the three surfaces marked in color and then click on the command "Fillet" to create the 1 mm edge roundings.

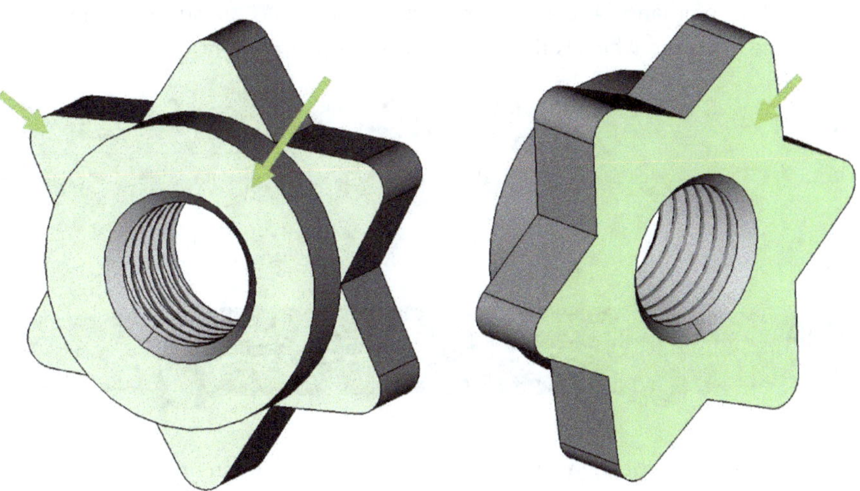

Lastly, we can assign the material with the command "Appearances". As with the dumbbell bar, I opt for "Chrome".

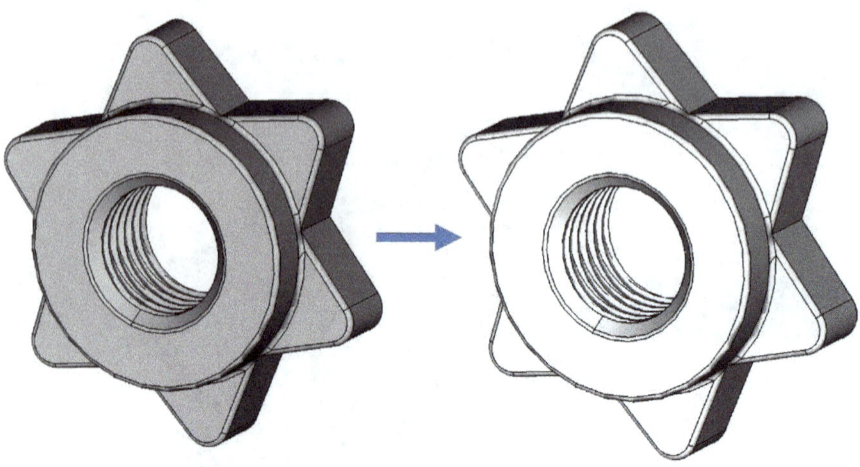

Excellent, now that we've constructed all the components we need, let's turn our attention to assembly in the next chapter and then enjoy our great construction!

5.5 Assembling the dumbbell set

For the assembly, we start in a new document and switch to the workspace "A2Plus" ①. Before we can insert parts, we first have to save the file. Then we click on the command "Add a part from an external file" ② and select the barbell bar as the first part, which is then fixed in the 3D environment.

Before we can mount the weight plates, we need to take an <u>important</u> step. We open the file for the 10 kg weight plate first (open as normal; without a command), select the two folders ① in the structure tree that contain the objects for the lettering, right-click on them and select "Delete" ②. It is critical that we click on "No" ③ in each of the <u>two</u> following pop-up windows, where we are asked whether we also want to delete the contents of the folders. We do not, but only intend to delete the folders but keep their contents. We have to do this as otherwise an error would occur during assembly. If you only delete the folders, this should not change lettering at all. We then do the same with the 5 kg weight plate.

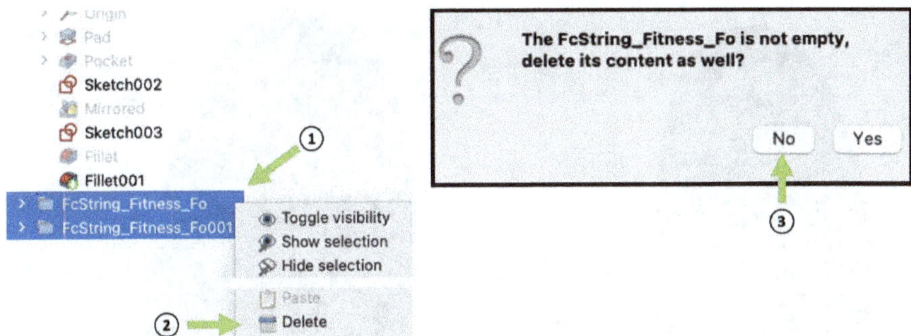

When done, close the weight plate files and switch to the "Assembly" file, which already contains the dumbbell bar. Here we execute the command "Add a part from an external file" ① and load a 10 kg weight plate into the file next. We then link the axes of the two components concentrically by clicking in the hole of the weight plate ② (hold down the CTRL key), the outer edge of the barbell bar ③ and the command "Add AxisCoincident constraint" ④ one after the other. Confirm with "Accept".

With the second constraint, we define the final position of the weight plate. To do this, we click on the following surfaces ① and ② (hold down the CTRL key) and then on the command "Add PlaneCoincident constraint" ③.

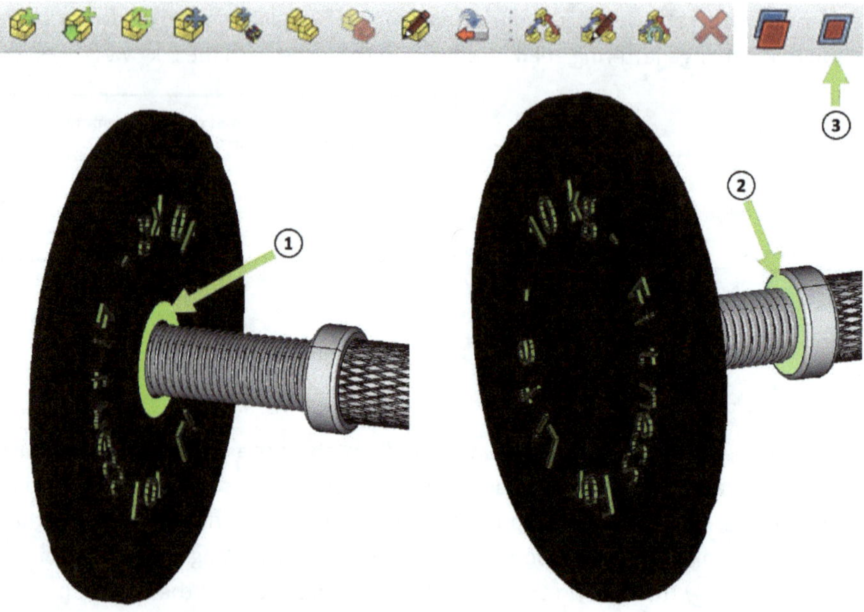

Perfect. We do the same procedure with a second 10 kg weight plate on the opposite side. To do this, simply repeat all the previous steps from the command "Add a part from an external file". This gives us the following intermediate result.

Now we also mount an additional 5 kg weight plate on each side. The assembly process is identical to that for the 10 kg weight plates. The only difference is that we select the outer attachment ② of the 10 kg weight plate — instead of the dumbbell bar — in the "Add PlaneCoincident constraint" command.

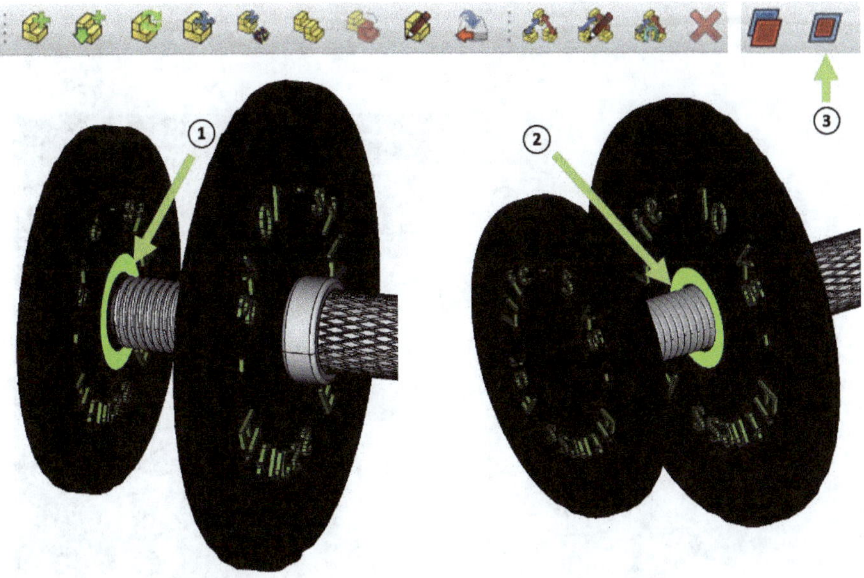

By now, you should have mounted the dumbbell bar with two 10 kg and two 5 kg weight plates.

Now it's time to fit both star nuts. We can also assemble these in the same way as before. To achieve this, we select the outer edges ① and ② of the two parts for the concentric connection of the two axes, for example. For the congruent connection, we select the inner surface ④ and the outer attachment ⑤, as we did for the other parts.

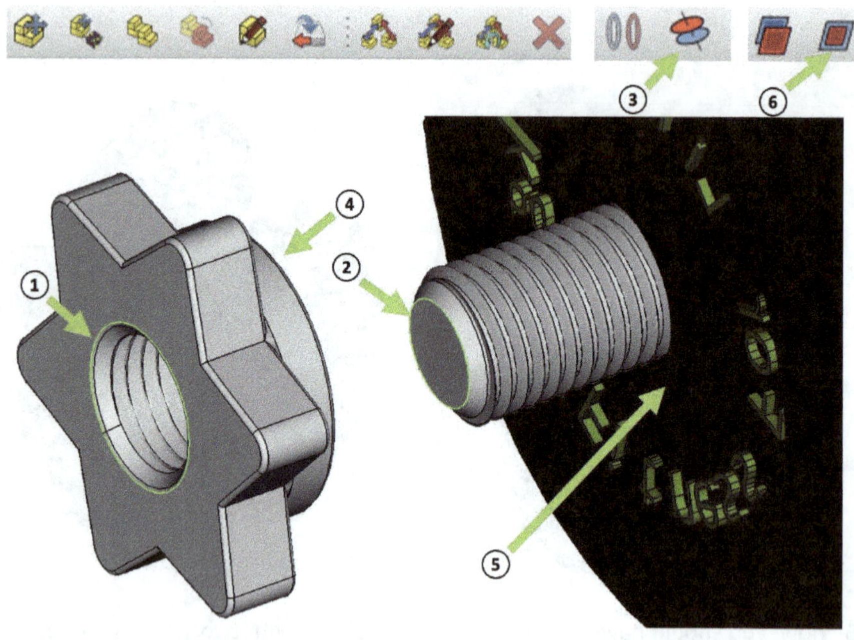

Once we have also fitted the second star nut on the opposite side, we can finally save the project. Excellent, we've done it! Great that you stuck with it to the end.

Closing words

Congratulations! You have successfully completed the advanced course.

You have now taken your CAD skills to a new level by familiarizing yourself with the design of complex assemblies. Together, we have designed four challenging objects, learning new functions and consolidating the use of familiar "FreeCAD" features. Be proud of your achievements!

What now?

There will probably be another part of this book series in which you can find the construction of other complex parts and assemblies. Please visit my author page on Amazon regularly to stay informed about new publications, especially if you want to further develop your skills and tackle more projects. Whether there will be another part also depends on you and your feedback. Feel free to leave a short positive review, it's very easy and quick.

3D printing — making your designs a reality

If you would now like to create your digital models physically, I recommend you take a look at 3D printing. Holding your designs in your hands creates a great feeling and also has great practical benefits (e.g., spare parts, individual objects). Take a look at my course *"3D printing | step by step"*, which explains how to get started with this fascinating technology.

Your opinion counts!

If you enjoyed this course, I would be delighted to receive a review and some feedback. Your review will help other readers to decide, and also help me to better tailor future courses and books to your needs. Thank you very much for your support!

Please take a brief look at the overview of my books on the following pages. Perhaps you will discover one or more exciting titles. See you soon and good luck with your future design projects!

Books on topics you might also like

All books are available online on the usual sales platforms. It's best to just search for the title, or feel free to visit my author page. Some of the books may not be published yet and will be released or found soon. Take a look at the books of your choice and your copy as e-book or paperback!

3D Printing:

CAD, FEM, CAM (3D Object Creation, Design, Simulation):

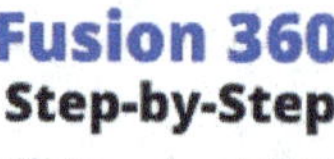

Electrical Engineering:

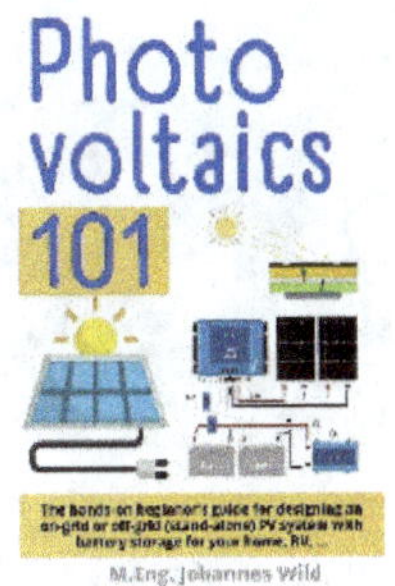

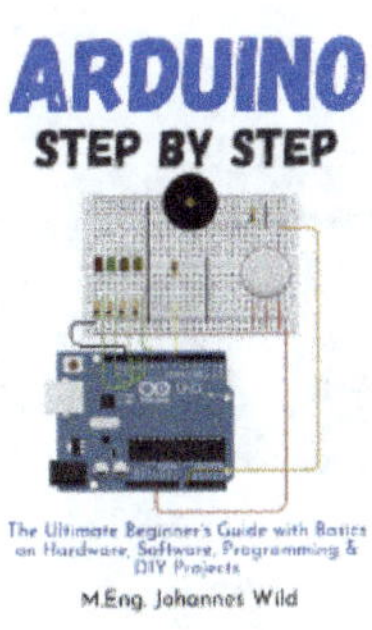

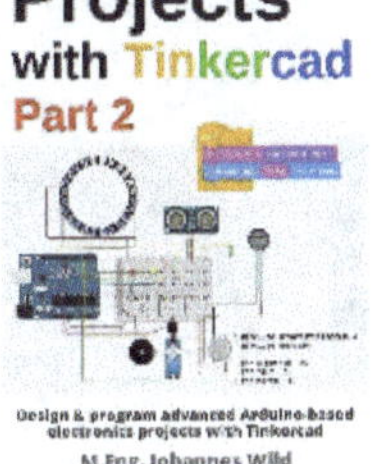

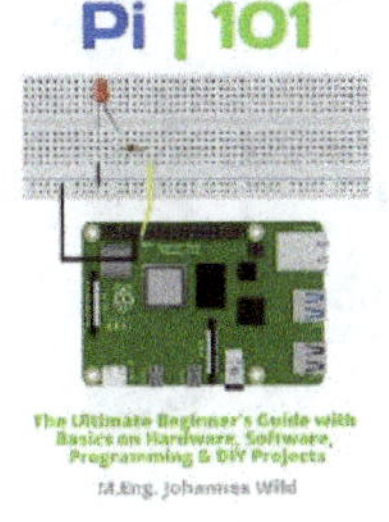

Programming and other Software:

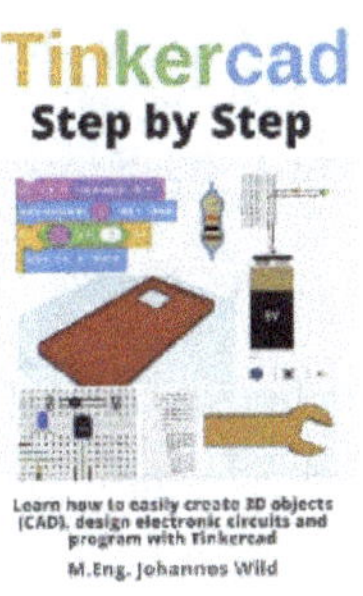

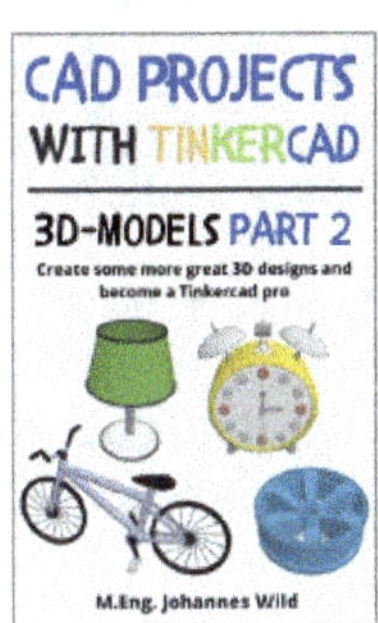

There are also identical video courses for some of these books:

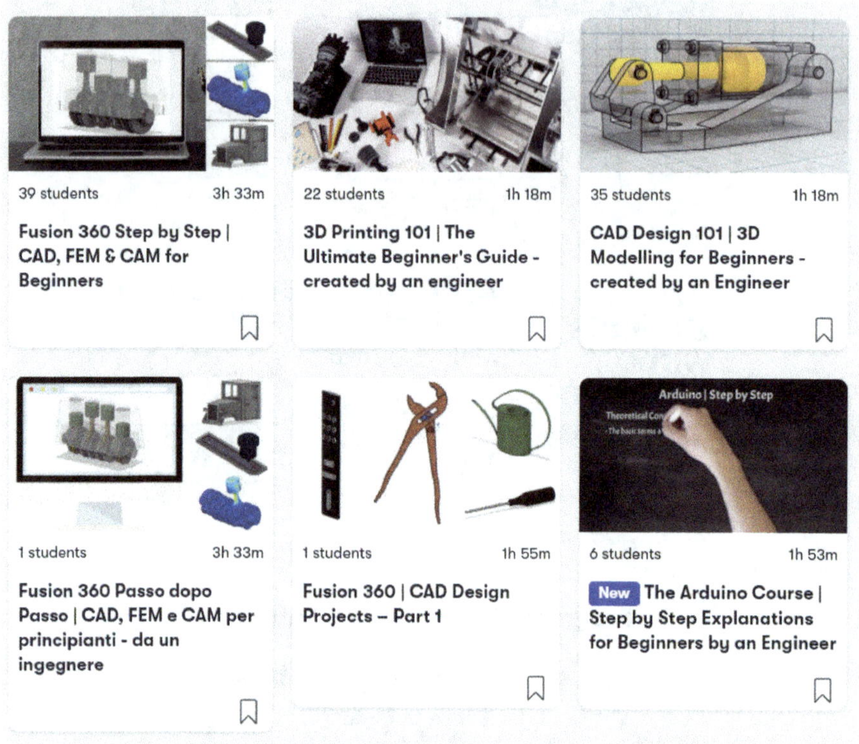

They are hosted on the learning website: skillshare.com

Be sure to use my following friends & family referral link to get a month of membership for free!

(I will get a little bonus if you choose to stay, so we will be both happy. Thanks in advance!)

https://www.skillshare.com/r/profile/Johannes-Wild/854541251

It is best to copy the link in your browser to access the free month!

Sign up today and deepen your knowledge!

Imprint of the author / publisher

© 2024

Johannes Wild
c/o RA Matutis
Berliner Straße 57
14467 Potsdam
Germany

Email: 3dtech@gmx.de

This work is protected by copyright

Thank you so much for choosing this book!